Fluorescent Probes

Based on a meeting "Fluorescent Probes in Proteins and Membranes" held at the Royal Institution of Great Britain, Thursday 8 November 1979.

Fluorescent Probes

Edited by

G. S. Beddard and M. A. West

The Royal Institution of Great Britain
Albemarle Street, London, England

 ACADEMIC PRESS · 1981

A Subsidiary of Harcourt Brace Jovanovich, Publishers

London New York Toronto Sydney San Francisco

ACADEMIC PRESS INC. (LONDON) LTD.
24/28 Oval Road,
London NW1

United States Edition published by
ACADEMIC PRESS INC.
111 Fifth Avenue
New York, New York 10003

British Library Cataloguing in Publication Data
Fluorescent probes
  1. Cell membranes
  2. Fluorescence spectroscopy
  I. Beddard, G. S.   II. West, M. A.
  574.87′5   QH601

ISBN 0-12-084680-2

Printed in Great Britain by
Thomson Litho Ltd, East Kilbride, Scotland

CONTRIBUTORS

BEDDARD, G.S.   *Davy Faraday Laboratory of the Royal Institution, 21 Albemarle Street, London W1X 4BS, UK*

BISBY, R.H.   *Department of Biochemistry, University of Salford, Salford M5 4WT, UK*

BRAND, L.   *Department of Biology and McCollum-Pratt Institute, The Johns Hopkins University, Baltimore, MA 21218, USA*

CONSTABLE, D.   *Department of Biochemistry, University of Birmingham, B15 2TT, UK*

COSSINS, A.R.   *Department of Zoology, University of Liverpool, Liverpool L69 3BX, UK*

CUNDALL, R.B.   *Department of Biochemistry, University of Salford, Salford M5 4WT, UK*

DALE, R.E.   *Paterson Laboratories, Christie Hospital and Holt Radium Institute, Manchester M20 9BX, UK*

DAVENPORT, L.   *Department of Biochemistry, University of Salford, Salford M5 4WT, UK*

EDIDIN, M.   *Department of Biology and McCollum-Pratt Institute, The Johns Hopkins University, Baltimore, MA 21218, USA*

JOHNSON, I.D.   *Department of Biochemistry, University of Salford, Salford M5 4WT, UK*

JOHNSON, S.M.   *Institut für Strahlenchemie, Max-Planck-Institut für Kohlenforschung, Stiftstrasse 34-36, D-4330, Mulheim a.d. Ruhr, West Germany*

MUNRO, I.H.   *Daresbury Laboratory, Warrington, Lancs. WA4 4AD, UK*

NOVROS, J.   *National Health Laboratories, 1007 Electric Avenue, Vienna, VA 22180, USA*

ROTH, S.   *Department of Biology and McCollum-Pratt Institute, The Johns Hopkins University, Baltimore, MA 21218, USA*

TEALE, F.W.J.   *Department of Biochemistry, University of Birmingham, Birmingham B15 2TT, UK*

THOMAS, E.W.   *Department of Biochemistry, University of Salford, Salford M5 4WT, UK*

THULBORN, K.R.   *Department of Biochemistry, University of Oxford, South Parks Road, Oxford OX1 3QU, UK*

PREFACE

The photophysical and photochemical properties of probes in
biological and in model systems are topics of intense current
research. The wavelength, intensity, time and polarisation
dependence of fluorescence probes can provide information on
properties as diverse as electric fields or changes in
conformation and may be used in the formidable task of
establishing structure-function relationships in proteins
and cell membranes.
The wide scope of interactions which can be studied by
these fluorescent molecules is illustrated in the articles
in this volume which are based on talks, although much
extended and updated, given at a meeting on 'Fluorescence
Probes in Proteins and Membranes' held in the Royal
Institution in November 1979.
The dynamic properties of fluorescence probes have
depended on the development of fast (nanosecond and picosecond) kinetic techniques which are described, together
with several applications, in articles by Cossins, Munro
and Beddard. Cossins used phase fluorometry to study the
ability of cells to modulate membrane fluidity in the face
of perturbations induced by external stimuli such as changes
of temperature. Time resolved fluorescence polarisation
spectroscopy has been used by Munro to observe the site and
temperature dependence of the rotational reorientation of
tryptophan residues in proteins. Synchrotron radiation was
used as a pulsed light source to excite the single tryptophan
residues. Beddard described the motion of probes bound to
the oligosaccharide groups on glycophorin in erythrocyte
membranes and outlines the techniques for picosecond spectroscopy. Time-resolved fluorescence anisotropy was used by
Johnson to compare non-tumourgenic and tumourgenic cell
membranes and Bisby *et al.* described the unusual photophysical properties of several diphenyl hexatriene derivatives and their application to the association of urushiol
(poison ivy allergen) in lipid membranes.
Although dynamic measurements of probes now often super-
cedes those made under steady state conditions, these latter

PREFACE

measurements still provide an accurate comparative index of
a number of properties as illustrated in articles by Thulborn,
Dale *et al.* and Teale and Constable. Thulborn compares the
steady-state fluorescence polarisation properties of
N-(9-anthroyloxy) fatty acid probes which report from
different transverse locations within the hydrocarbon region
of the lipid bilayer. Energy transfer from the mitogenic
lectin, concanavalin-A on the surface of normal and malignant cells was used by Dale and coworkers to probe the
general distribution of this lectin on cell surfaces. Teale
and Constable describe the preparation and application of
several unsymmetrical disulphide reagents for thiol-specific
fluorescent labelling to macromolecules.

A substantial bibliography is appended which covers
publications on fluorescence probes and their use in
biological model systems. The emphasis, in the latter years,
was directed towards applications to proteins and membranes.
We hope this bibliography will provide a ready source of
information for newcomers and indicate developments in
related fields for those research workers and students
already studying fluorescence probes.

April 1981
                           Godfrey S. Beddard
                           Michael A. West

CONTENTS

INTRAMOLECULARLY-QUENCHED
THIOL-SPECIFIC FLUORESCENCE PROBES

F.W.J. TEALE and D. CONSTABLE*

Department of Biochemistry
University of Birmingham
Birmingham B15 2TT

INTRODUCTION

The usefulness of fluorophors attached covalently to the surface of macromolecules is widely recognised (Kapoor, 1977; Kanoaka, 1977) and some of the information potentially available is outlined in table 1.

TABLE 1
Information potentially available from fluorescent probes

Probe environment	Complexing
Distances	Orientations
Accessibility	Rotations
Conformation change	Lateral diffusion
Concentrations	Microviscosity
Electric fields	Group reactivities
pH	Functional correlations

Although in any particular system the information is of a low-resolution kind, by using alternative fluorophors and all the fluorescence parameters summarized in table 2, discrete information can often be resolved. Although some physical applications, such as overall shape estimation through

---

* Present address : Miles Laboratories Ltd., Bridgend

TABLE 2
Experimental parameters and information content

Wavelength (λ)	Probe environment, lateral diffusion, electric field, pH
Intensity (Q)	Complexing, accessibility, distance, pH, reactivities, concentrations
Time (τ,ρ)	Distance, rotations, size, heterogeneity
Polarisation (P,A)	Rotation, orientation, microviscosity, change in conformation, size

anisotropy measurements require a random surface labelling, most investigations are best carried out by unique site labelling, since specific structural insights can only be hoped for by placing the fluorescent probe into a definite location and, if possible, orientation in the macromolecule. This can be done in two general ways :- (a) using the known specificity requirements of the protein binding site (or enzyme catalytic or control site) to complex selectively with the reagent or (b) using a 'specific' group reagent to pick out the only, or alternatively the most reactive, member of a particular class of reactive side-chain. These two approaches have both been usefully employed. Technique (a) presupposes a knowledge of the site specificity and reactive groups and can be used for assay of activity or for site titration, depending on the stability of the adduct. Technique (b) assumes strict specificity, which is often invalidated because of the notoriously diverse environments which a particular side-chain may experience on the macro-molecular surface. The use of several different 'specific' reagents can safeguard against, or reveal, aberrant labelling reactions. In both techniques an important aspect is the ease of following site reaction. It is very convenient if more than one experimental spectroscopic parameter can be used to quantitate the extent of reaction. One approach is to construct linked fluorophors which become separated during the site-labelling reaction. If resonance energy transfer (RET) occurs between chromophores, then reaction will be accompanied by the appearance of donor emission and simultaneous decrease in sensitized acceptor fluorescence. Thus emission measurements at two wavelengths will characterise the amount

of reaction. This idea was first put forward by Stryer in
1966 who suggested enzyme reactions of the structure
Donor-b_1-b_2-b_3-Acceptor in which any one linkage needs to be
cleaved to release donor fluorescence. Haughland and
Stryer (1967) subsequently synthesised the reagent shown in
figure 1c, anthraniloyl-p-nitro phenolate, to titrate the
active sites of trypsin and α-chymotrypsin. Here the acceptor
is non-fluorescent, as is the intact reagent, whereas the
acylated enzyme is blue fluorescent and the released p-nitro-
phenol ionises to the yellow phenolate anion. Thus two
signals, fluorescence and absorbance changes, accompany the
catalysed reaction, from the donor and acceptor moities
respectively. For enzyme assay purposes, the site specificity requirements can often be built into the chain of linkages, while the terminal chromophores can be any chemically
suitable RET pair with sufficient transfer efficiency
relative to the extreme chromophore separation. This strategy
has been used by Katchalski and coworkers (Yaron, Carmel and
Katchalski-Katzir, 1979) to construct a range of fluorogenic
substrates for several specific enzymes (figure 1a, 1b). In
two of the reagents shown (d,f) the quenching of the anthraniloyl donor is by collisional, rather than RET, interactions
with the nitro-benzyl terminus. In the NAD^+ analogue shown
(figure 1e) interaction between ε-AMP and nicotinamide cations
effectively suppresses the fluorescence of the former (Barrio
et al., 1972). The same idea can be applied to specific
group reagents if reaction can be coupled to chromophore
separation. An obvious candidate for this strategy is
revealed by inspecting some of the established thiol-specific
reagents shown in figure 2 (Brocklehurst, 1979). Here the
disulphide-thiol exchange type lends itself conveniently to
the construction of chromophore pairs. Moreover this
particular inherently reversible reaction is probably the most
selective and specific for thiol. Bearing in mind the fact
that in macromolecules thiol groups are often comparatively
few in number and often functionally important, we have
constructed a range of unsymmetrical disulphide reagents
(UDR) shown in figure 3, in which the fluorescent donor
moiety is quenched by RET to a generally non-fluorescent
acceptor linked together by the disulphide bridge. Which
side of the reagent links itself to the macromolecular thiol
and which chromophore is released into the medium depends on
the chemical nature of the linkages. The aliphatic moiety
attaches itself exclusively to the protein thiol, and the
aromatic thiol is released into the medium. This mechanism
can be understood either as a consequence of the difference
in the intrinsic redox potentials of the reactants and
products, or as consequence of resonance stabilisation in

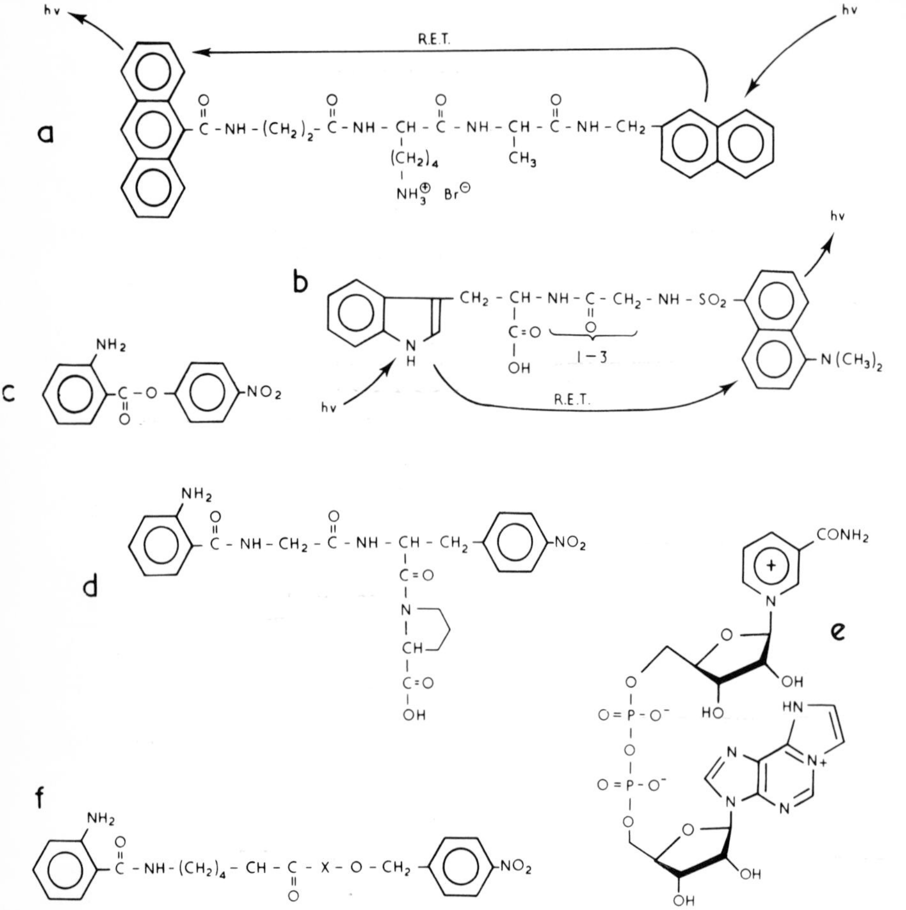

Fig. 1. Intramolecularly-quenched fluorogenic reagents

the aromatic thiol-thiolate equilibrium.

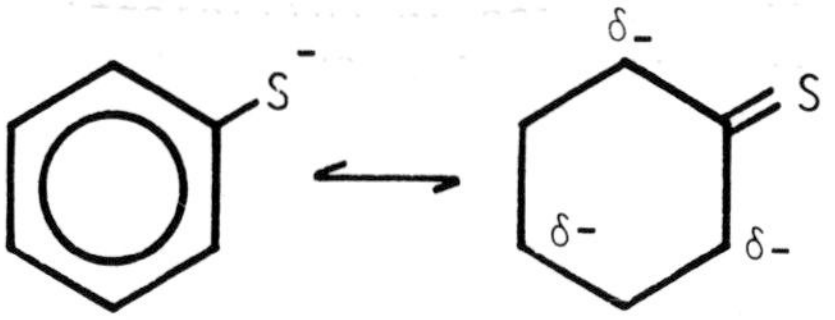

Of two possible arrangements shown in figure 3, the second type, in which the fluorescent moiety is released into the medium, although difficult to synthesize, is very useful for the assay of thiol in those pigmented proteins which have intense light absorption. Whereas the differential increase

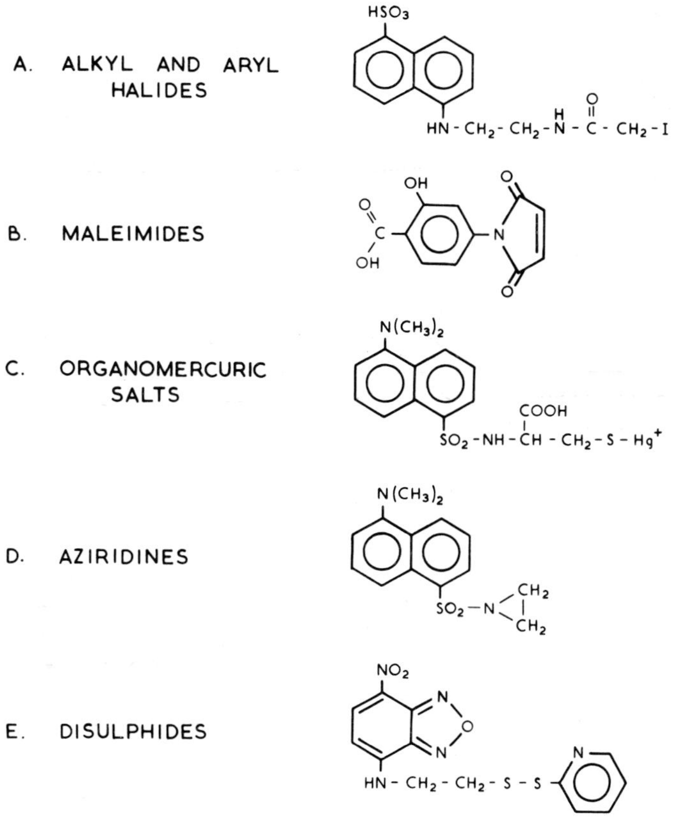

A. ALKYL AND ARYL HALIDES

B. MALEIMIDES

C. ORGANOMERCURIC SALTS

D. AZIRIDINES

E. DISULPHIDES

Fig. 2. *Some examples of the main types of 'thiol-specific' reagents used for fluorescence labelling*

in absorbance at 413 nm produced by reaction of DTNB* with haemoglobin is very difficult to measure against the background of the heme Soret band, the presence of released fluorophor is easily measured in sufficiently dilute solutions. Needless to say, the alternative fluorophor-labelled

*Abbreviations :

AEDANS	Acetyl N'-(5-sulphonic-1-naphthyl) ethylene diamine
DEAE	Diethylaminoethyl
DTNB	Dithiobisnitrobenzoate (Ellman's Reagent)
NAC	N-acetyl cysteine
RET	Resonance energy transfer
UDR	Unsymmetrical disulphide reagent

Protein + Fluorescence Label — Chromophore release

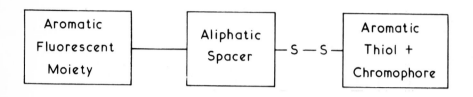

Fluorescent moiety release — Chromophore onto Protein

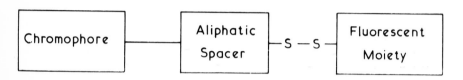

Fig. 3. The two arrangements of unsymmetrical disulphide reagents for thiol labelling

haemoglobin shows negligible signal because of efficient RET to heme. In both UDR reagents the disulphide cleaved fluorophor, or absorber, is released into the medium in which the quantum yield or molar absorbance, respectively, can be determined absolutely. Thus the extent of reaction can always be easily determined, without assumptions about the quantum yield or molar absorbance in the reaction site. Especially in the case of fluorochrome labelled proteins, comparison of the fluorescent intensity data with the released chromophore enables the quantum yield, and hence the site heterogeneity to be monitored as reaction proceeds. A simple illustration of some possibilities is given in figure 4, in terms of the reactivities and quantum yield modification. Besides the thiol-disulphide exchange type some other side-chain specific reagents lend themselves to the two chromophore arrangement. The amino-specific imidate can be synthesized so that the R_1 and R_2 groups in the intact imidate have a quenching inter- action which ceases on reaction with the amine R_3NH_2.

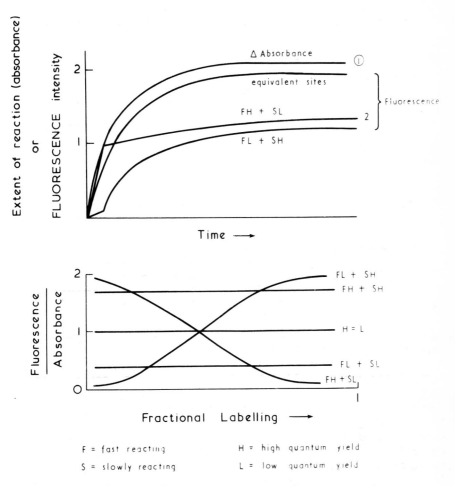

Fig. 4. Thiol-site heterogeneity as reflected in the quantum yield of the attached probe and reaction rate

Acid anhydrides can also be constructed in a similar way.
In conclusion it may be said that in principle reagents can be synthesized for the assay of any enzyme which catalyses bond cleavage between groups suitable for the attachment of the chromophores. Quenching can be brought about through collisions or complex formation, rather than RET. No overlap of chromophore spectra is necessary in the choice of the quenching group. A disadvantage of these alternative reagents is that fluorescence can appear by other processes than cleavage of the linkage. Adsorption of the intact reagent on to protein sites so that quenching collisions or complexing are prevented could be mistaken for reaction with thiol.

CHOICE OF CHROMOPHORES, THE LINKAGE, AND SYNTHESIS

Most of our work has been done with the chromophores shown
in table 3, although many more have been investigated. The
fluorophor donors were chosen to cover a wide range between
the blue and yellow spectral regions. Several trade-offs and
compromises were necessary. Small chromophore size was felt
to be important, and imposed limits on one or two aromatic
rings. A spectral window through which to excite protein
aromatic amino-acid residues was also desirable to maximise
the ease of RET measurements. Reasonable solubility in
aqueous media was also important and indeed it was found that
insolubility was often associated with an unacceptably high
rate of disproportionation to the symmetrical disulphides.
Some otherwise useful chromophores were found to adsorb onto
protein surfaces so complicating fluorimetry. The most
useful acceptor group was one half of DTNB, although others
have been used (Ellman, 1959) to obtain absorbance at longer
wavelengths. Synthesis of the fluorescent thiol was carried
out by standard techniques, often by chemical or electrolytic
reduction of the symmetrical disulphide. The thiol was then
reacted with an excess of the aromatic disulphide. The excess
unreacted disulphide and the aromatic thiol were separated
from the UDR by solvent extraction, ion exchange or gel
filtration, depending on the particular side-groups present.
In some cases, the fluorescent moiety was added last to
facilitate reduction to the thiol. Other techniques for UDR
synthesis, such as the thiosulphonate pathway gave unacceptably
low yields compared with that described above. Preparations
were normally carried out on the 100 mg scale, but this could
easily be increased without difficulty. By modifying the
acceptor acid side chain, reagents of different overall change
were produced in an attempt to investigate the importance of
the thiol site polarity. In all cases, the residual quantum
yield of the UDR was less than 2% of the cleaved components
and stable as dry solid for many months in a deep freeze. In
aqueous solution at room temperature there was no photo-
sensitivity, but a very slow dismutation to the symmetrical
disulphides became measurable after one day at neutral pH,
increasing markedly in alkaline solutions. Reactivity toward
aliphatic thiol was up to two orders of magnitude faster than
shown by DTNB, perhaps due to a lower redox potential or
possibly due to the presence of the aliphatic spacing residues
in the linkage decreasing steric hindrance.

SPECIFICITY FOR THIOL

Each UDR was incubated with a concentrated ($\sim$ 1.0 M) aqueous solution of analogues of all the protein side-chain groups, except thiol and with proteins known to have no thiol groups, such as lysozyme and pepsin. No reaction, judged by the appearance of fluorescence or colour change occurred despite many days exposure at 5°C. Only in 2M neutral hydroxylamine or ethylenediamine was some cleavage noted, possibly owing to reductive side-reactions. Rapid reaction was demonstrated with mercaptoethanol, cysteine, N-acetyl cysteine, thioglycollate, dithionite, sulphite and dithiothreitol as expected.

WHAT IS THE QUENCHING MECHANISM IN THESE UDR's ?

Several processes could lead to the non-fluorescence of the potential fluorophor in these compounds. Three obvious possibilities are static complexing, collisional complexing and RET. The first possibility was made unlikely by the near perfect coincidence of the UDR absorption spectrum in 'good' solvents with an equimolar mixture of the separate appropriate analogues of the two moieties. Strong complexing of the terminal chromophores could thus be ruled out. In some cases, 'poor' solvents could be found in which demonstrable complexing took place. This was very marked in bis-anthraniloyl cystamine, which had poor solubility in water, a low quantum yield and perturbed absorption spectrum. In contrast, ethanolic solutions had a quantum yield five times higher, comparable with the free fluorophor, and unperturbed absorption. The residual fluorescence of each UDR was unchanged in a range of good solvents of graded dielectric constant extending from dioxan to formamide, again suggesting that little intramolecular complexing is present. It is, of course, very difficult to exclude a process of weak ground state complexing, which increases greatly in the S_1 excited state. Complexing between the fluorophor and the disulphide bridge as a cause of quenching was excluded by synthesis of the UDR analogues in which the acceptor chromophore was replaced by an aliphatic group, and in the other the disulphide link was replaced by two methylenes. In the former the fluorescent yield approached that of the free fluorophor, while in the latter analogue fluorescent quenching was virtually complete. The collisional quenching process is characterised by dynamic time-dependence and should be inhibited in highly viscous media. Comparison of the UDR dissolved in methanol and propylene glycol at 293 K, where these otherwise similar solvents have a forty-fold viscosity ratio, revealed no significant changes in quantum yield, and

TABLE 3

Donor and acceptor fluorphors

DONOR	COMMENT	ACCEPTOR	COMMENT
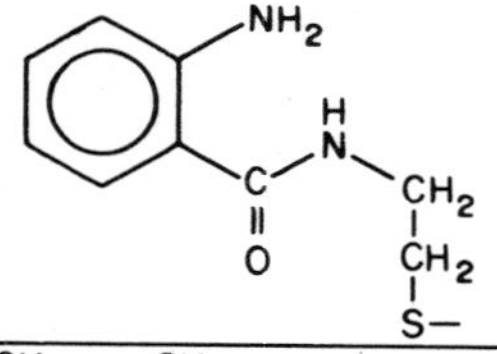 (structure: benzene ring with NH$_2$, C=O, N–H, –CH$_2$–CH$_2$–S–)	Blue emission at 412 nm Solvent sensitive	(structure: benzene ring with COOH, NO$_2$, –S–)	Pale yellow max 325 nm Negative charge
		(structure: benzene ring with CO·N·CH$_2$·CH$_2$OH, H, NO$_2$, –S–)	Pale yellow max 325 nm Neutral
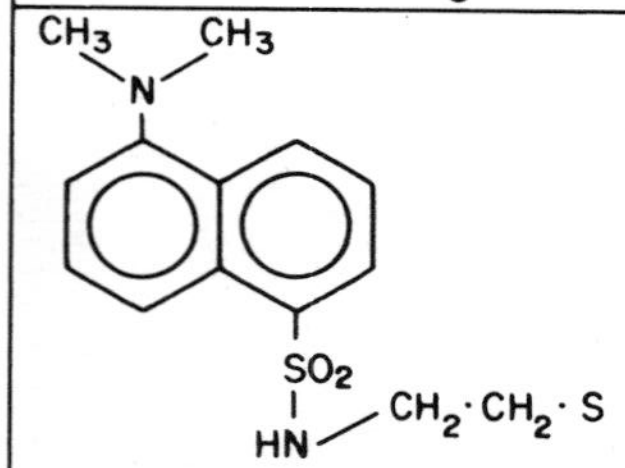 (structure: naphthalene with CH$_3$–N–CH$_3$, SO$_2$, CH$_2$·CH$_2$·S, HN)	Yellow emission 540 nm in water Solvent sensitive	(structure: benzene ring with CH$_2$–N$^{(+)}$(CH$_3$)$_3$, NO$_2$, –S–)	Pale yellow max 325 nm Positive charge
		(structure: benzene ring with H$_2$N, NO$_2$, –S–)	Yellow, extended absorption Neutral
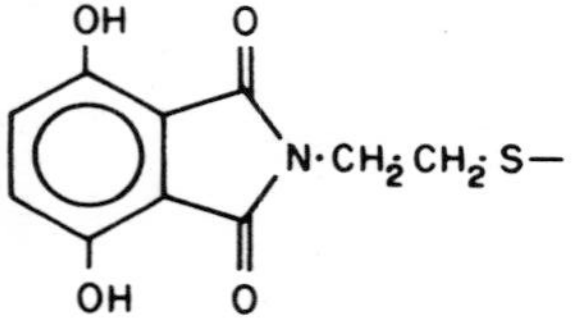 (structure: ring with OH, OH, O=, =O, N·CH$_2$CH$_2$S–)	Yellow emission pH and redox sensitive pK$_a$'s 7 and 9	(structure: benzothiazolium with NO$_2$, CH$_3$, N$^{(+)}$, S, –S–)	Yellow, extended absorption Positive charge

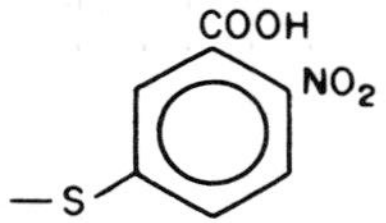

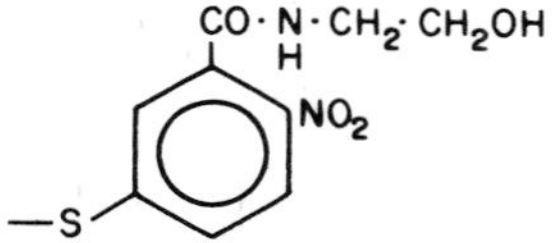

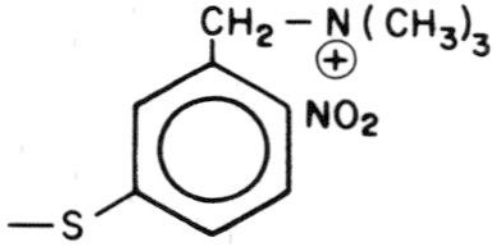

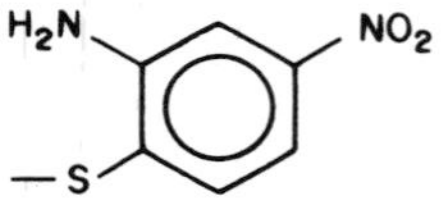

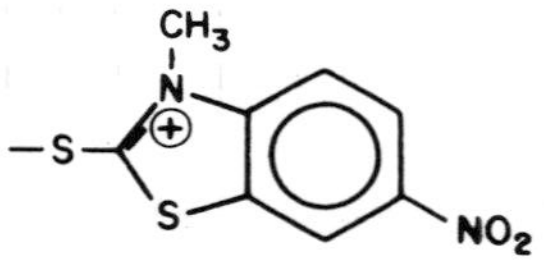

the same observation applied on cooling both solutions to
200 K, at which the glycol becomes a rigid glass. The same
negative result was given by incorporation of the UDR into
dry films of polyvinyl alcohol. It was concluded that
either collisional quenching was unimportant, or improbably,
was replaced by static complexing in rigid media. The effect
of further temperature reduction was significant, however.
Cooling to 70 K in liquid nitrogen caused a large increase
in fluorescence, accompanied by phosphorescence, especially
in UDR such as the DNS compound with a very small overlap
integral between fluorescence spectrum and acceptor absorption.
This low temperature effect can be explained by reference to
figure 5.

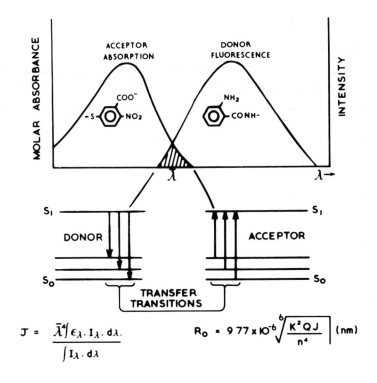

$$J = \frac{\bar{\lambda}^4 \int \epsilon_\lambda \cdot I_\lambda \cdot d\lambda}{\int I_\lambda \cdot d\lambda} \qquad R_0 = 9\,77 \times 10^{-6} \sqrt[6]{\frac{K^2 Q J}{n^4}} \quad (nm)$$

*Fig. 5. Energy levels of the donor and acceptor chromophores,
showing the origin of J, the overlap integral*

Both chromophores in the UDR have similar absorption spectra
so that the usual diagnostic test for RET, donor sensitized
acceptor emission, would be excluded even if the acceptor
was fluorescent. The overlap transitions which permit RET
are thermal in origin, and the relatively small overlap

integral is adequate for efficient transfer provided the interchromophore separation is sufficiently small. At very close range, very small overlap values sustain quantitative RET, until exchange reactions take over as the important transfer process. At sufficiently low temperatures the thermal depletion of the vibrational manifold causes virtual cessation of RET, and the quantum yield approaches that of the free donor. This reduction in the overlap absorbance can be observed directly in DTNB or a UDR at low temperature. It was noteworthy that none of the UDR probes became fluorescent when bound without cleavage to serum albumin, since the terminal chromophores were made to have similar or opposite overall charges. Since complexing and collision frequency might be expected to be charge-dependent, these observations once again supported RET as the important quenching processes. The quantum yield and lifetime parameters of two UDRs are shown in figure 6 and suggest that heterogeneity of structure exists, probably through different chain extensions and orientation effects.

DONOR	ACCEPTOR	$J(cm^6/mol)$	R_0 (nm)	F/F_0 (calc)	F/F_0 (obs)	τ/τ_0 (obs)
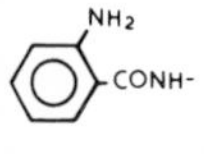 NH₂—⬡—CONH–	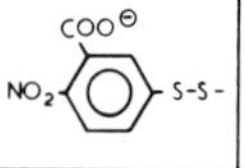 COO⊖ NO₂—⬡—S-S–	$4·6 \times 10^{-16}$	1·80	0·028*	0·01	0·20
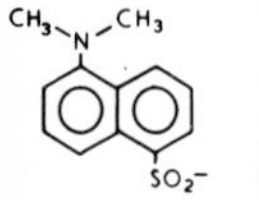 CH₃–N–CH₃ (naphthalene) SO₂⁻	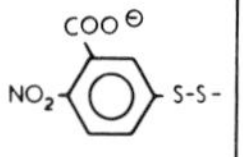 COO⊖ NO₂—⬡—S-S–	$1·6 \times 10^{-17}$	1·06	1·06	0·02	0·18

*Calculated for maximum chromophore separation, 1·0nm

Fig. 6. Energy transfer parameters and the observed fractional quantum yields and lifetimes of two UDR probes

It is noticeable that, particularly for the DNS compound, the observed fractional quantum yield is much smaller than that calculated for maximum chain extension and chromophore separation. This suggests that many molecules have a folded structure favouring energy transfer. This conclusion is supported by the high values of fractional lifetime compared with fractional quantum yield, diagnostic of fluorophor heterogeneity since lifetime is intensity weighted.

UDR REACTIONS WITH THIOL GROUPS

Some of the applications of UDR probes are shown in figure 7. For the moment, only three applications will be considered,

assay, inter-residue distances, and thiol-site polarity and
heterogeneity.

In the assay of aliphatic thiols, these reagents conferred
the increased sensitivity to be expected of the fluorescence
technique compared with absorption methods using DTNB and
similar disulphides. The reaction rate was generally greatly
increased compared with DTNB, typically by up to two orders
of magnitude. Comparison with conventional methods showed
excellent quantitative agreement. By carrying out the assay
at relatively high concentrations at which the absorbance of
released aromatic thiolate could be accurately measured by
difference spectroscopy at 413 nm, the stoiciometric assay
could then be carried out fluorimetrically at concentrations
as low as 1 nanomolar in favourable systems. Protein reactivity studies were initiated with mammalian species variants
of serum albumins, characterised by a single thiol surface
group in a molecular mass of 65000 Daltons, and having a
common overall structure. Two species, human and rabbit,
contain a single tryptophan residue, while two others, sheep
and bovine, contain two tryptophans, and all four have a
similar large (∿ 20) number of tyrosine residues. All the
UDR probes reacted rapidly and stoiciometrically to yield
stable fluorescence-labelled albumin with blue-shifted
emission spectrum characteristic of a 'hydrophobic' location.

1. Thiol assay generally

2. Protein thiol environment, polarity, exposure

3. Mobility of attached probe, anisotropy

4. Protein thiol reactivity and heterogeneity

5. Distance between chromophores

       TYR/TRP →   PROBE

       PROBE   →   PROBE

       PROBE   →   Intrinsic chromophore

6. Excited state equilibria, pK*'s

7. Electron scavenging

8. Conformational change

Fig. 7. Some possible applications of UDR probes

Typical excitation and emission spectra are shown in figure
8 together with the tentative domain structure proposed by
Brown (1974) and Geisow and Beaven (1977). Surprisingly,
there was no transfer of tryptophan to thiol-attached probe

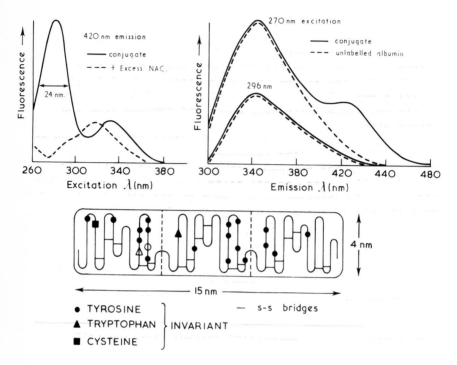

Fig. 8. Excitation and emission spectra of human mercaptalbumin thiol-labelled with anthraniloyl UDR, and the tentative domain structure of this protein

in any of the four albumins, presumably due to the relative remoteness of the fluorescent tryptophan from the thiol site or just possibly an unfavourable orientation. In contrast relatively efficient transfer (~ 30%) from tyrosine to probe was indicated by the small bandwidth (24 nm) of the sensitized excitation spectrum, characteristic of tyrosine. For trypto-phan, a bandwidth of 30 nm would be expected. Inspection of the tentative structure suggests that about 20% of the total tyrosine residues are within the R_0 characteristic distance for tyrosine-probe transfer, and others more remote may contribute, *via* intertyrosine transfer, to probe excitation. The essential similarity of the RET parameters in one or two tryptophan albumins suggests that both tryptophan locations are beyond effective transfer range, or that the non-invariant second tryptophan location has a low quantum yield or unfavourable orientation. These observations contrast with the relatively good tryptophan transfer reported in albumin thiol-labelled with AEDANS (Hudson and Weber, 1973). The different behaviour of the two probe types requires

further investigation. In these experiments, care was taken
to prepare albumins with strictly one mol thiol per macromolecule, from material with much lower thiol content. The
albumin thiol was first activated with excess N-acetyl
cystine (NAC) and then passed down a DEAE column or a
suitable affinity column, the most useful material being an
agarose-p-aminophenyl mercuric chloride. The bound mercapto-
albumin was eluted with NAC, which was removed from the
protein with Dowex 2 ion-exchange resin. All the albumins
were found to show non-specific reversible binding of up to
two mols of UDR, causing tryptophan quenching. This adsorption was prevented by competitive binding of fatty acids at
this adsorption site, or gel filtration. There were marked
differences in reaction rates shown by the albumins towards
UDR differing in overall charge. By using the anthraniloyl
UDR probes of different acceptor charge shown in table 3 to
react with the thiol site of human mercaptalbumin , the data
shown in table 4 was obtained.

TABLE 4
Charge environment of human mercaptalbumin thiol

Probe charge	Normalised initial reaction rate			
	pH 6 low salt	pH 6 high salt	pH 4 low salt	5 mol/ℓ G HCl
+1	1	8	1	2
0	3	8	2	9
-1	10	10	10	10

The rate differences, and the effect of pH and neutral salt
concentration suggested a positively charged environment for
the thiol site, possibly resulting from tertiary folding,
and not seriously disturbed by the acid-induced N-F
conformational change. Although the trypophan and tyrosine
emission was dramatically changed at the N-F transition,
probe emission by direct excitation was hardly altered.
    Papain was chosen as an example of an active-site thiol
enzyme (Baines and Brocklehurst, 1978). The UDR probe
reacted rapidly and stoichiometrically with the single
active thiol to inhibit the enzyme. The emission was
characteristic of a 'hydrophobic' environment, but was
easily accessible to added collisional quencher. The
excitation spectrum (figure 9) showed that the equivalent
of one out of six tryptophan residues sensitizes the probe

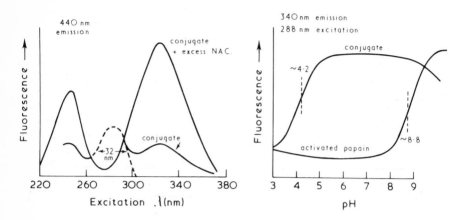

Fig. 9. Excitation spectrum of anthraniloyl papain and the tryptophan fluorescence – pH profiles of the conjugate and the native enzyme

emission. In the simplest terms, this indicates that only one of the six tryptophan residues is within transfer distance, and also that little intertryptophan transfer occurs. Also in figure 9 is shown the pH profile of the activated papain tryptophan fluorescence compared with that of the conjugate. The large shift in quenching profile may implicate active site thiol as the tryptophan quencher, which normally masks the acid quenching process presumably due to the titration of neighbouring carboxylate residues. Myosin was selected as a large multi-thiol enzyme. Very rapid reaction occurred with approximately four moles of UDR which inactivated ATPase activity and F-actin binding capability, while a total of about 30 moles reacted at longer exposures to excess reagent. The emission spectrum of successively reacting probes showed a progressive shift to longer wavelengths, and some emission depolarisation became measurable. This behaviour was interpreted as an initial rapid reaction at the catalytic sites in the globular head region, followed by progressively slower classes of site probably in the exposed helical portion of the fibrous protein.

HETEROGENEITY OF THIOL SITE

The simultaneous release of donor fluorescence and increased acceptor absorbance which accompany the reaction between UDR and thiol enables the relative average quantum yield of the probe site to be followed during the course of reaction. When several different classes of thiol site are present,

a plot of the ratio of intensity to absorbance increase
against time reveals heterogeneity of type as reflected in
quantum yield. Average lifetime measurements provide similar
information. Thus heterogeneity can be shown by reaction
rates, Stokes shift, quantum yield and fluorescence lifetime.
Differentiation between sites can be further investigated by
changes in pH and ionic strength, and also by external
quencher effects. In myosin, two groups of thiol were easily
demonstrated, and in membrane preparations the same differences were revealed. In rat mammary Golgi cell membrane a
class of fast reacting thiols in sites conferring blue-shifted
high quantum yield emission constituted about a third of the
total. The remaining thiols showed a broad spectrum of slower
reactivities, the least reactive being most red-shifted and
with lowest quantum yield. Treatment of the membrane with
neutral detergents altered these rather unreactive sites
towards higher reactivities and blue-shifted emissions.

SUMMARY

The unsymmetrical disulphide reagent (UDR) probes are
probably the most thiol-specific fluorescence labelling
reagents available for macromolecules. In many cases the
simultaneous provision of fluorescence and absorbance changes
with the exchange reaction permits facile measurement of the
extent of reaction without the usual recourse to external
fluorescence standards. In principle, a wide variety of
fluorophors can be incorporated into UDR probes since the
minimum overlap integral required is easily achieved.
Reversal of labelling by the presence or addition of excess
aliphatic thiol is at once an experimental convenience, and
a fundamental disadvantage of UDR labelling compared with
other irreversibly covalently attached. Nevertheless the
conjugates given by these latter reagents should preferably
be checked against those provided by the appropriate UDR in
order to ensure that thiol-specific labelling has been
achieved.

REFERENCES

Baines, B.S. and Brocklehurst, B. (1978), *Biochem. J.*
  173, 345-347
Barrio, J.R., Secrist, J.A. and Leonard, N.J. (1972), *Proc.
  Nat. Acad. Sci. U.S.A.*,
Brocklehurst, K. (1979), *Int. J. Biochem.*, 10, 259-274
Brown, J.R.(1974), *Fed. Proc. Fed. Am. Soc. Exp. Biol.*,
  33, 1389

Ellman, G.L. (1959),*Arch. Bioch. Biophys.* $\underline{82}$, 70-77

Geisow, M.J. and Beaven, G.H.(1977), *Biochem. J.*
$\underline{163}$, 477-484

Haugland, R.P. and Stryer, L. (1967)*In* "Conformation of Biopolymers". (G.N. Ramachandran, ed.), pp. 321-328. Academic Press, New York

Hudson, E.N. and Weber, G.(1973), *Biochemistry*, $\underline{12}$, 4154-4161

Kanaoka, Y. (1977), *Angew. Chem. Int. Ed. Engl.*, $\underline{16}$, 137-147

Kapoor, M. (1977), *J. Scient. Ind. Res.*, $\underline{36}$, 74-91

Stryer, L. (1966) *In* "Biology and Exploration of Mars", (C.S. Pittendrigh and W. Vishmar, eds), Nat. Acad. Sci., Washington, pp. 394-401

Yaron, A., Carmel, A. and Katchalski-Katzir, E. (1979), *Anal. Biochem.*, 95, 228-235 (and references therein)

DISCUSSION ON DR. TEALE'S PAPER

Dr. Tegmo-Larsson : Can you give an example of these
energy transfer reagents used to study changes in protein
conformation ?

Dr. Teale : In muscle protein there are well-established
interactions between small polypeptide fragments and the
myosin macromolecule in the globular head region. There
are strategically-placed thiol groups in the interacting
system which can be fluorescently labelled. We are hoping
to measure, through resonance energy transfer, functional
changes in geometry in this complex system.

Dr. Bayley : Does one observe probe migration from one
site to another ?

Dr. Teale : We have no evidence for this. When the probe
leaves it just goes into solution. Tests by differential
labelling with iodoacetamide before and after the disulphide
probe reaction give no evidence for changing probe location.

FAST TIME RESOLVED FLUORESCENCE MEASUREMENTS
AND THEIR APPLICATION TO SOME PROBLEMS
OF BIOLOGICAL INTEREST

GODFREY S. BEDDARD

*Davy Faraday Laboratory of the Royal Institution
21 Albemarle St., London W1X 4BS*

Time resolved fluorescence spectrometry has become an increasingly valuable technique in biological studies and the accessible time range has recently been extended about a thousand-fold by the use of mode-locked lasers. The fluorescence decays and anisotropic motion of probes in lipid bilayers (Kawato *et al.*, 1977), of intrinsic probes in proteins proteins (Beddard *et al.*, 1980; Grinvald and Steinberg, 1976), the process of energy transfer in photosynthetic organisms (Campillo *et al.*, 1977; Searle *et al.*, 1978; Beddard *et al.*, 1979) and in *Halobacterium Halobium* (Hirsch *et al.*, 1976; Alfano *et al.*, 1976) have all been studied by fast fluorescence techniques.

Four principal methods of utilising the ultra-short light pulses from mode-locked lasers have been used and these are (1) the optical Kerr effect, (2) the streak camera, (3) time-correlated single photon counting and (4) the sum-frequency light gate. Generally methods (1) and (2) use single pulses from a mode-locked solid-state laser and methods (3) and (4) are used with a continuously running mode-locked gas laser. All four methods have been used in our laboratory (Beddard *et al.*, 1979, 1980, 1981; Porter and West, 1978) although the optical Kerr method is now used rarely.

The relative advantages and disadvantages of the latter three methods will now be summarised :

(a) The Streak Camera
The streak camera is probably unique in time measuring instruments in that an image of the time-profile of the event (a laser pulse or fluorescence decay, for example) is obtained. The image is formed on a phosphor which is later

'read' either by a photographic plate or by an electron beam
in an optical multichannel analyser (OMA). The streak camera-
OMA combination is equivalent to a photomultiplier and oscill-
oscope as used for longer time scale measurements.
    The camera, when operated at its fastest streak speed has
a time resolution of $ca.$ 2 ps, although this is greatly reduced
at the streak speeds normally used to record weak fluorescence.
The time resolution is obtained by sweeping a voltage ramp
across a drift tube down which are travelling electrons
generated by photoelectron emission from a photocathode. The
photoelectrons are generated from the light pulse viewed
through a narrow (10-300 μm) slit in front of the photocathode.
The time resolution is determined by the slitwidth (since an
image is obtained on the phosphor) and by the sweep speed of
the voltage ramp. The first photoelectrons passing down the
tube are deflected to one side but as time progresses the
electrons emitted from later in the light pulse are deflected
to the other side of the tube, the deflection being linear
with time. Thus a spatial displacement is achieved which
represents a time profile. The intensity of the image on the
phosphor is proportional to the intensity of the light input
and horizontal displacement (mm/ps) is proportional to time.
In order to achieve the fastest time response, a narrow slit
is needed for the input light. This is copied on to the
output phosphor and unlike other time measuring instruments,
the time resolution becomes worse as the time scale increases
i.e. fewer mm/ps. This arises because the slit width on the
phosphor is always the same length irrespective of the time
scale. Thus for long times (fluorescence decays greater
than 0.5 ns) the streak camera has only comparable or worse
time resolution than photon counting. It is best used for
the 1-300 ps range of lifetimes and is particularly good in
resolving emission risetimes.
    Our neodymium-glass laser-streak camera system, outlined
in figure 1 in block form, has been extensively used in the
study of photosynthetic systems (Porter et al., 1976; Searle
et al., 1978) and particularly in the study of energy transfer between intrinsic chromoproteins in the red alga
Porphyridium cruentum (Porter et al., 1978; Searle et al.,
1978). In this alga an accessory light gathering pigment-
protein complex grows on the face of the thylakoid membrane
in which the photosynthetic apparatus is contained. This
accessory pigment complex, a phycobilisome, contains three
main pigment types, each based on a tetrapyrrole chromophore,
which are phycoerythrin, phycocyanin and allophycocyanin.
A structural model of this chromoprotein consists of an
allophycocyanin (APC) core surrounded by R-phycocyanin (RPC)
with phycoerythrin (PE) on the periphery, figure 2 (Gantt

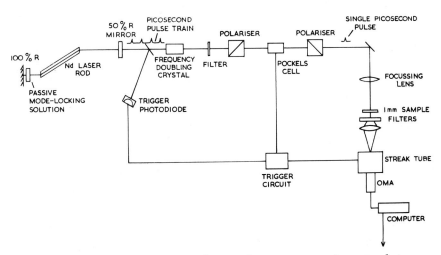

Fig. 1.   Neodymium laser and streak camera system used to measure fluorescence lifetimes

et al., 1976).   The APC channels energy into the chlorophyll in the thylakoid membrane.   In decreasing order, the energy levels of these pigments are, PE > RPC > APC > chlorophyll and Forster energy transfer which transfers energy from one pigment to the next also follows this order even though transfer between pigments of the same type is also possible. Since the phycobilisome is large, a prolate hemiellipsoid 30 nm high and 45 nm wide, and even though Forster transfer between between individual pigments is very fast, finite

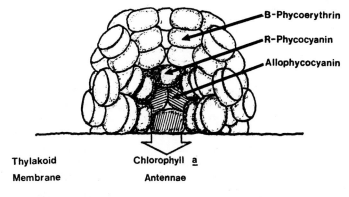

Fig. 2.   Model of a phycobilisome as envisioned to exist in the alga, Porphyridium cruentum, based on Gantt et al., Biochim. Biophys. Acta (1976), 430, 375.

rise-times have been observed in the pigments RPC, APC and
chlorophyll which are not directly excited by the 530 nm
laser pulse. The phycoeryhthrin is excited and this fluorescence rises promptly with the laser pulse and decays in a
complex way but with a 1/e lifetime of 70 ps. The decays for
the other pigments RPC, APC and chlorophyll are 90, 118 and
175 ps, respectively, with fluorescence rise-times of 12,
24 and 50 ps. Figure 3 compares the duration of the laser
pulse with the decay of PE and the appearance of the fluorescence of APC. The distinction in time of these decays is
only possible because of their differing fluorescence maxima
which allows selection of one species almost to the exclusion
of the others.

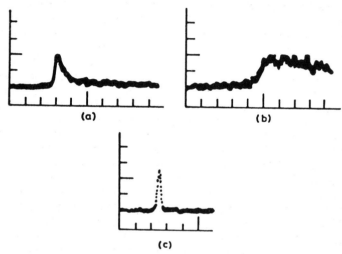

(a)   (b)

(c)

*Fig. 3.   Energy transfer from phycoerythrin to allophycocyanin
in phycobilisomes : (a) phycoerythrin fluorescence (b) allophycocyanin fluorescence (c) laser pulse profile. (a) and (b)
time scale 120 ps/div. (c) time scale 60 ps/div.*

The single shot method used with the streak camera requires
a high energy per pulse and this is often a disadvantage since
it may introduce unwanted effects such as excited state
annihiliation (Swenberg *et al.*, 1976; Porter *et al.*, 1977),
stimulated emission (Hindman *et al.*, 1978) or sample decomp-
osition during the measurement (Fleming *et al.*, 1978). If the
laser pulse energy is lowered, the precision of measurement
is impaired. Even at low pulse energies of 10^{14} photons/cm^2/
pulse considerably more light enters the sample than from a
laser used for photon counting. A synchronously operated
streak camera in combination with a mode-locked CW dye laser
may prove the solution to these problems (Welford *et al.*, 1980)

In spite of these problems, the streak camera is particularly useful for measuring fluorescence rise-times as a result of energy transfer or intramolecular complex formation (Eisenthal 1975) in the time range up to 400 ps. The instrument provides high data acquisition rates (depending on the repetition rate of the laser) and degradation can generally be avoided by flowing light sensitive materials through the sample cell. Species with short chemical lifetimes can be studied, which is not always possible using a repetitive laser and low data acquisition rates.

(b) Time-correlated single photon counting
The next technique we consider, that of single photon counting provides another solution to the problem of fast decay time measurements with a time resolution of about 50 ps. In this case the resolution is limited by transit time fluctuations in the photomultipliers, walk jitter in the constant fraction discriminators, linearity in the time to amplitude converters and also the resolution in the analogue to digital converters in the multichannel analyser. Even though this method has the poorest time resolution of the three, it has an extremely high precision and is probably the most widely used.
    The instrument used is outlined in figure 4 and has been described in detail elsewhere (Beddard *et al.*, 1980). An argon ion laser (Coherent Radiation) is mode-locked at 514 nm to produce 80-100 ps FWHM pulses at 1.2 W average power. These pulses are used to excite a sodium fluorescein extended cavity jet-stream dye laser producing 4 ps pulses at 550 nm and at 95 MHz. The pulse repetition rate was reduced to 45 kHz using a repetitively pulsed Pockels cell placed between crossed polarisers and triggered by countdown logic from the RF source used to mode-lock the ion laser. A contrast ratio, between transmitted and rejected pulses of more than 200 : 1 is obtained. A cavity dumper can also be used to reduce the pulse repetition rate. Fluorescence is detected by means of a Philips XP2020Q photomultiplier tube and scattered light is removed by filters. Vertically polarised excitation was selected by a calcite polariser and the fluorescence collected at right angles through polarisers set vertically (intensity at time t, $I_{||}$ (t)) and horizontally ($I_{\perp}$ (t)). Rotation-free decays were measured with the polariser set at 54.7°. The emission anisotropy r(t) was calculated from the expression :

$$r(t) \quad = \quad \frac{I_{||} (t).G \; + \; I_{\perp} (t)}{I_{||} (t).G \; + \; 2I_{\perp} (t)} \qquad (1)$$

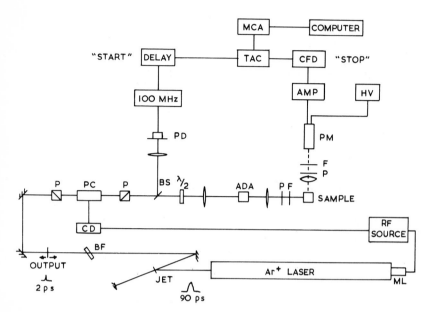

Fig. 4. Schematic of the photon counting fluorescence lifetime instrument. ML – mode-locker on the argon ion laser; JET – jet-stream dye laser; BF – birefringent wavelength tuning element; OUTPUT – dye laser output mirror on a translation stage; P – polarisers; CD – countdown   ; PC – Pockels cell driven by pulses from the mode-locker via CD ; BS – beam-splitter; ADA – frequency-doubling crystal; PD – timing photodiode; F – filter; PM – photomultiplier; AMP – amplifier; CFD – constant fraction discriminator; TAC – time to amplitude converter; HV – photomultiplier power supply.

The time resolution limitation, as has been noted, is due to jitter in the electronic components alone since the dye laser used produces pulses less than 4 ps long and with negligible jitter. If the total jitter from all sources is τ_m, this is derived from a Gaussian combination of broadened responses by $\tau_m{}^2 = \Sigma(\tau_j)^2$ for individual responses, τ_j. Table 1 shows a comparison of the responses for various photomultipliers and electronic components. Because of the relationship between τ_m and τ_j it is possible to use a wide laser pulse, say 100 ps from an ion laser, as an excitation source without seriously degrading the system response. This is particularly convenient with molecules such as bilirubin (Tran and Beddard, unpublished work) which absorbs in the 400-500 nm region where there are ion laser lines that can be mode-locked but where it is difficult to generate light from a dye laser.

TABLE 1

*Estimated contributions to the measured
instrument response times*

Component	FWHM jitter (ps)
BPW28A photodiode	<50
XP2020Q photomultiplier	150-190
Varian VPM152 photo-multiplier	<200
Philips 56TUVP/56DUVP photomultiplier	575-600
Ortec 436 100 MHz discriminator*	210
Ortec 473A CFD	<50
Ortec 473A CFD**	250-300
Ortec 463 CFD**	190-220
Ortec 934 CFD**	∿150
RCA C31034 photomultiplier	∿150-190

* For the photodiode pulse height distribution
** For the XP2020 pulse height distribution
CFD = constant fraction discriminator

The following example discusses the application of the photon counting fluorescence instrument to study the motion of eosin probes bound to glycophorin in erythrocytes. A vaiety of biologically important phenomena involve molecular interactions at the cell surface. Oligosaccharide side chains of glycoproteins and glycolipids present in the plasma membrane typically cover the surface of eukaryotic cells and these moities are frequently implicated in surface recognition events (Nicolson, 1976). Of particular interest has been the band 3 complex in human erythrocyte cell membrane since this and some associated proteins (such as glycophorin) are understood to play a role in anion transport through the membrane (Hughes, 1975; Nicolson, 1976). Band 3 and glyco-phorin span the erythrocyte membrane and have oligosaccharide chains on the outside of the cell surface. On the cytoplasmic face of the membrane are smaller proteins such as spectrin which with actin provide a peripheral network responsible for the biconcave cell shape (Nicolson, 1976).

Transient dichroism experiments (Nigg and Cherry, 1979; Cherry, 1979) of eosin-5-maleate labelled band 3 show that the minimum functional unit of the protein is a stable dimer of 180000 Daltons. The evidence for this comes from the

diffusion coefficient D for rotation about an axis perpendicular to the membrane surface. This diffusion constant
was 1296 $\pm$ 99 s^{-1} but was 1492 + 202 s^{-1} when the band 3 was
chemically dimerised. If band $\overline{3}$ has a cylindrical shape, a
radius of 4 nm and a height of 4 nm would be consistent with
the data. The glycophorin polypeptide chain contains about
140 amino acids. About 60% by weight of the protein is
carbohydrate mainly as two principle types of oligosaccharide
which bear receptors for lectins and viruses. Glycophorin is
proposed to be amphipathic with a hydrophilic-hydrophobic-
hydrophilic structure. A 23 amino acid sequence devoid of
charged amino acid residues, predominantly composed of
hydrophobic residues anchors the protein to the membrane.
Figure 5 shows a possible arrangement of band 3 and glycophorin in the membrane.

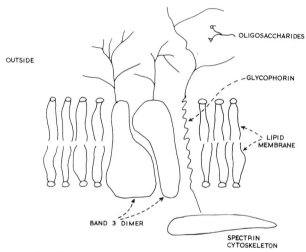

*Fig. 5. Hypothetical model of the human erythrocyte band 3
glycophorin complex*

In the present study, sialic acid and galactose residues
on the erythrocyte surface were oxidised and eosin-5-thio-
semicarbazide (ETSC) conjugated to the resultant aldehyde
(Cherry *et al.*, 1976, 1980). After labelling the sialic
acid, the membranes contained 1-2 µg eosin/mg membrane
protein, corresponding to about 10^6 molecules of eosin per
cell. The results of sodium dodecylsulphate/polyacrylamide
gel electrophoresis on the labelled membranes showed the
eosin fluorescence to be on the monomer and dimer bands of
glycophorin-A. To demonstrate that band 3 (which forms a
protein complex with glycophorin) was not labelled this
protein was dimerised by Cu-phenanthroline catalysed disulphide

bridge formation and no change occurred in the position of the eosin fluorescence in the gel. The galactose-labelled membranes contained a similar ETSC content to those labelled at the sialic acid sites but in contrast to these latter, very diffuse fluorescence bands were observed in the gels, making it difficult to assign the label to individual bands. Some band 3 and lipid was probably labelled also.

Figures 6 and 7 show the time dependence of the eosin emission anisotropy calculated from the measured fluorescence decay curves and $I_{\parallel}$ (t) and $I_{\perp}$ (t). Because of the short duration of the instrumental response compared to the anisotropy, convolution of the anisotropy was not necessary. For both the sialic acid and galactose-labelled membranes, the anisotropy exhibited an initial decay reaching a constant level at longer times. Fitting the anisotropy to the equation (Jahnig, 1979) :

$$r(t) = (r_0 - r_\infty) \exp(-t/\phi) + r_\infty \qquad (2)$$

yielded ϕ = 3.8 ns for sialic acid and 2.9 ns for galactose labelled membranes. To check if faster motions are present, the anisotropy was taken at 9 ps/point (figure 6) and no detectable motion occurred in the range 0.2-1.0 ns. In addition to eosin fluorescence, the dichroism due to triplet-triplet absorption (Cherry,1979) has been measured on the 0.02-2.0 ms time scale. The limiting anisotropy was 0.02 + 0.01 for sialic acid labelling but was remarkably different for galactose-labelled membranes, reaching a maximum of 0.15 in the 1-2 ms range. The results from both fluorescence and triplet absorption studies are summarised in table 2.

TABLE 2

Time-resolved measurements on eosin-TSC-labelled erythrocyte membranes

	Sialic acid	Galactose
r_0	0.3 + 0.01	0.3 + 0.1
r_∞	0.22	0.22
ϕ	3.8 ns	2.9 ns
r_{ss}	0.26 + 0.01	0.29 + 0.01
r_{td}	0.02 + 0.01	0.15

r_{ss} = steady-state anisotropy; r_{td} = anisotropy measured by transient dichroism (time scale ms).
eosin/glycerol -20°C, r_{ss} = 0.33 + 0.01
eosin/water,ϕ = 550 ps

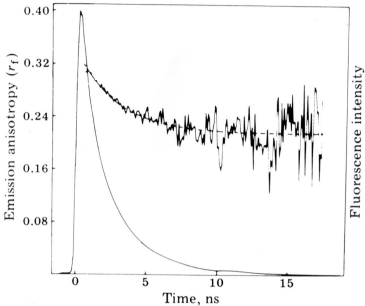

Fig. 6. *Decay of the fluorescence anisotropy of eosin-TSC-labelled galactose in erythrocyte membranes. The dotted line shows the fitted curve (from equation 2) to the experimental data, the smooth solid curve shows the fluorescence decay on the same scale. Zero time is set at the peak of the fluorescence since deconvolution is not necessary.*

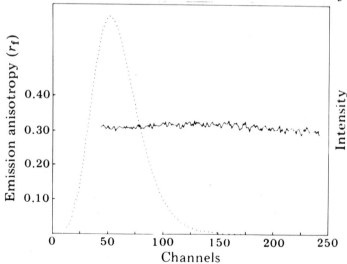

Fig. 7. *Fluorescence emission anisotropy as in fig. 6 but taken at 9 ps/point. The dotted line is the instrumental response curve.*

The anisotropy measured in all these experiments is typical
of restricted motion of the probe. The rapid decay of the
anisotropy within a few nanoseconds is due to localised
independent motion of eosin and a short segment of the oligosaccharide chain and is motion within a limited angular
amplitude. The order parameter, $s^2 = r_\infty / r_0$ obtained from
curve fitting is 0.8-0.9. A rigid system has an order
parameter of 1 while a system exhibiting unrestricted Brownian
motion has this parameter as zero. In sialic acid-labelled
membranes, the order parameter on the ms time scale is 0.2-
0.3 which implies a large amplitude motion reduces the
anisotropy from its high value at short times. The situation
in the galactose-labelled samples is more complex and it is
likely that the amplitude of the slower motions varies among
different binding sites. Since the anisotropy is still large
in the ms time range the motion of these sites must be
highly restricted (Cherry *et al.*, 1980).

(c) The sum-frequency light gate
The third technique to be considered in the sum-frequency
light gate (Mahr and Hirsch, 1975; Hirsch and Mahr,1979).
This is a direct optical, as opposed to electronic, method
of obtaining fluorescence decay curves. The method uses
parametric sum-frequency generation in an anisotropic crystal
Frequency-doubling, as often used with solid state lasers,
is a degenerate example of this process. A photon of
frequency ω_3 is generated from a photon of the strong pump
beam of frequency ω_1 in its interaction with a weaker beam
of photons of frequency ω_2 i.e.

$$\omega_3 = \omega_2 + \omega_1$$

If ω_1 is the laser pulse and ω_2 the fluorescence frequencies
respectively, the fluorescence is gated by the laser (assumed
to be of shorter duration than the fluorescence) since ω_3
can be generated only when ω_1 and ω_2 are present at the
same time in the crystal.
    An instrument for measuring fluorescence decays by this
method (Beddard *et al.*, 1981) is shown in figure 8 and is
one of a class of pump-probe instruments which can be used
to perform, with slight modifications, picosecond absorption
and rotational diffusional measurements (Shank, 1980;
Beddard and Westby, 1981). The laser input beam is split
into two, one part travelling down a variable delay line
whose length is controlled by a translation stage driven by
a stepping motor. The other part of the beam is used to
excite the fluorescence which together with the laser light

from the variable arm is focussed into an angle-tuned
frequency summing crystal. The power per unit length of the
sum frequency light $P_3(L)$ for a crystal of length L is given
by :

$$P_3(L) \simeq g. \; L^2 \; P_1 P_2 \; \omega_3/\omega_1$$

where g is a constant which depends on the frequencies and
optical properties of the crystal. If the (laser) pump beam
is swept along the fluorescence decay profile, the signal
intensity at the sum frequency is mapped out in time as the
optical delay line changes its length. The time resolution
depends only on the laser pulse length (assuming a thin
crystal is used for frequency conversion) since the delay
line can be incremented by as little as 0.1 ps (or less if
needed) per data point. The sum frequency is passed to a
UV-transmitting, visible-blocking filter and hence to a monochromator. A photomultiplier is used for fluorescence
detection and photon counting used to obtain a signal count
fcllowing subtraction of background counts. This signal is
stored in one channel of a multichannel analyser and the
experiment repeated for a different position of the delay
line. The total fluorescence intensity can be monitored and
used to normalise the signal in each channel for fluctuations
in the laser intensity.

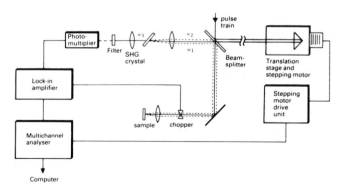

*Fig. 8. Schematic diagram of up-conversion arrangement for
picosecond fluorescence measurements*

Figure 9 shows the recorded fluorescence decay profile of
the dye cresyl violet in acetone solution together with the
laser autocorrelation profile. The double exponential fit
to the data gives a rotational relaxation time of 78 ± 4 ps
in agreement with other measurements (Millar *et al.*, 1979)

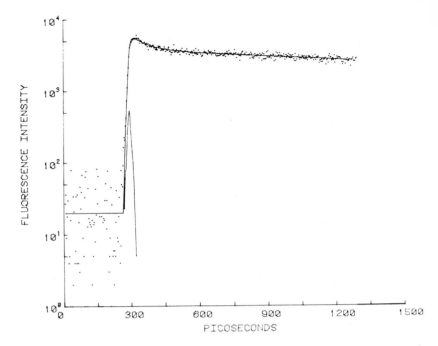

Fig. 9. Fluorescence of cresyl violet in acetone measured by the sum-frequency light gate with excitation and emission polarisers parallel. The initially decaying portion of the curve shows the effect of rotational diffusion. The smooth line is the fitted curve to the experimental points after convolution of two exponentials with the laser autocorrelation profile the lower narrow solid line. Time scale - 3 ps/ch.

The fluorescence intensity I(t) at time t after excitation is given by

$$I(t) = \int_{-\infty}^{t} F(t') \, L(t-t') \, dt' \qquad (4)$$

where F(t) is the true molecular fluorescence and L(t) the laser pulse shape. The signal detected at the crystal is

$$S(t) = \int_{-\infty}^{t} I(t') \, L(t-t') \, dt' \qquad (5)$$

and combining these two equations, we find that if A(t) is the laser pulse autocorrelation function, then S(t) is given by :

$$S(t) = \int_{-\infty}^{t} A(t') \, F(t-t') \, dt' \qquad (6)$$

The autocorrelation of the laser pulse is measured by removing the sample and adding the two laser pulse frequencies together in the crystal. Using this autocorrelation is an advantage when measuring short decay times since no approximations are necessary on the relationship of the actual pulse shape to its autocorrelation.

The particular frequency selected from the fluorescence at a particular crystal angle depends on the phase matching conditions. Altering the angle of the crystal to the input beam selects a different range of frequencies. In this way time-resolved spectra can be generated by scanning the fluorescence wavelength selected at some fixed time after excitation.

Figure 10 shows the effect of photoselection when the fluorescence is detected in the parallel and perpendicular directions after excitation in the vertical direction. The effects of Brownian motion on the fluorescence was clearly resolvable in the propanol solvent used for this experiment.

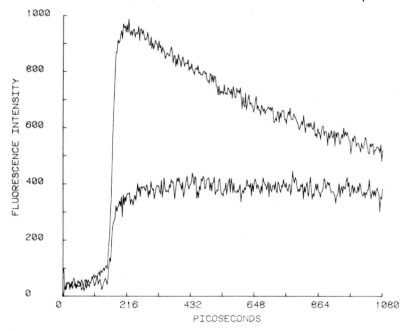

Fig. 10. Fluorescence profiles of cresyl violet measured in the parallel (top curve) and perpendicular (lower curve) directions to the excitation polarisation.

A rotational correlation time of 696 ± 15 ps was measured and this time can be calculated from the Debye-Einstein theory of rotational diffusion in both the 'stick' and 'slip'

limits. In 'stick', the solvent impedes rotational motion
sufficiently that motion in a liquid is considerably slower
than inertial motion (such as that occurring in the gas phase).
In the 'slip' limit, no solute-solvent interaction impedes
motion but the necessity of the non-spherical solute to
displace solvent occurs as it rotates. The calculated line
of rotational correlation with solvent viscosity in the
'stick' limit has a gradient of 100 ps/cpoise for cresyl
violet (compared to an experimental line of 300 ps/cpoise).
Hydrogen bonding is presumed to occur in the alcohol solvents
with the $-NH_2$ groups thus slowing down rotational motion.
No such bonding occurs with oxazine, in comparison, which is
symmetrical and has $-NEt_2$ in place of $-NH_2$ groups. Transient
dichroism experiments (Fleming et al., 1981) showed a
gradient of rotational lifetime with solvent viscosity of
87 ps/cpoise. Thus the inhibition of hydrogen bonding by
the bulky ethyl groups causes motion nearer to the slip limit
behaviour (60 ps/cpoise) than to the stick limit.
     The application of this method to the measurement of fluor-
escent probes in membranes and proteins does not yet appear
to have been attempted but should prove of value in measuring
rapid molecular motion where convolution processes in photon
counting lead to ambiguities. Additionally, since sum
frequencies are generated, fluorescence in the far red region
of the spectrum can easily be detected with a UV-sensitive
photomultiplier. Similarly difference frequency measurements
can also be made where fluorescence occurs in the UV by using
a red pump beam.
     Fluorescence in the far red from $Halobacterium$ $Halobium$
has been detected by Hirsch et al., (1976) using this light
gate method. The fluorescent species is bacteriorhodopsin
and the individual protein complexes contain a pigment
similar to the visual pigment found in the vertibrate retina.
This bacteriorhodopsin spans the bacterial membrane and during
the light driven cycle H^+ is ejected on the outside of the
membrane and regained from the inside. This proton pump
mediates in the formation of ATP which is used to fuel
metabolism. The chromophore is retinal bound to an opsin-
like protein by a protonated Schiff's base on the ϵ-amino
group of lysine. On excitation, the chromophore present in
the all-trans configuration undergoes a series of transform-
ations involving protonation and deprotonation of the Schiff's
base and conformational changes in the protein. Each stage
can be followed by its characteristic absorption spectrum
but when dark adapted the pigment is in its initial state in
the photochemical cycle. Thus excitation into this state
should yield information on the first photochemical or
photophysical process. In elegant experiments by Hirsch and

coworkers (1976) the fluorescence from the bacteriorhodopsin at 780 nm was shown to appear instantly with the laser pulse and decayed with a lifetime of only 15 $\pm$ 3 ps. The emission quantum yield was measured as 1.2-2.5×10^{-4} and the intrinsic lifetime was calculated to be 6 ns leading to an expected decay time of 140 fs - approximately 100 times shorter than the measured decay time. This emission does not originate from the ^1B_u state of the polyene chromophore but possibly from the ^1A_g forbidden state proposed to exist in polyenes. Since the fluorescence appeared with the laser pulse, this indicates that the fluorescence appeared from the excited pigment and not from one of its subsequent products in the photochemical cycle. Furthermore since the fluorescence decays with a 15 ps lifetime, the next species should appear with this same lifetime. In picosecond absorption experiments, however, the absorption characteristic of the second species in the photochemical cycle appeared in less than 1 ps. The interpretation of these contrasting observations has yet to be resolved.

CONCLUSION

The high time resolution and also high sensitivity of the light gate method should enable it to be used to study fluorescence probes. Probes such as eosin or ethidium bromide , which absorb in the 500-600 nm region of the spectrum appear attractive and fast motions (less than 10 ps) such as those suggested by Millar (1980) to occur in labelled DNA and RNA would be resolvable. Similarly intrinsic probes such as tryptophan or tyrosine where previous studies (e.g. Munro, this volume) have indicated fluorescence anisotropy in the subnanosecond region, are ameanable to study by the light gate method. The sum frequency method removes any distortion caused by instrumental effects and allows the anisotropy to be measured directly.

ACKNOWLEDGEMENTS

G.S.B. thanks the Science Research Council for financial support and acknowledges the collaboration of Dr. R. Cherry (ETH, Zurich) and Mr. T. Doust and Mrs. M. Westby (Davy Faraday Research Laboratory).

REFERENCES

Alfano,R.R., Govindjee, B., Becher, B. and Ebrey T.(1976) *Biophys. J.*, <u>16</u>, 1399

Beddard, G.S., Fleming, G.R., Porter, G., Searle, G. and
Synowiec,J.A. (1979) *Biochem. Biophys. Acta* <u>16</u>, 1399
Beddard, G.S., Fleming, G.R., Porter, G. and Robbins, R.
(1980)*Phil. Trans. Roy. Soc.*, <u>545</u>, 165
Beddard, G.S., Doust, T.A.M., and Porter, G. (1981)
submitted to Chem. Phys. Letters
Beddard, G.S. and Westby, M. (1981) *Chem. Phys.*in press
Campillo, A., Hyer, R., Monger, T., Parson, W. and Shapiro,
S. (1977) *Proc. Natl. Acad. Sci. U.S.A.*, <u>74</u>, 1977
Cherry, R.J., Burkli,A., Busslinger, M., Schneider, G. and
Parish, G. (1976) *Nature* <u>263</u>, 389
Cherry, R.J. (1979) *Biochem. Biophys. Acta* <u>559</u>, 289
Cherry, R.J., Nigg, E., and Beddard, G.S., (1980) *Proc. Natl.
Acad. Sci. U.S.A.*, <u>77</u>, 5899
Eisenthal, K.B. (1975) *Acc. Chem. Research,* <u>8</u>, 118
Fleming, G.R., Waldeck, D. and Beddard, G.S., (1981)
Il. Nouvo Chemie, in press
Fleming, G.R., Morris, J., Robbins, R. Woolfe,G., Thistle-
thwaite, P. and Robinson, G.W. (1978) *Proc. Natl. Acad.
Sci. U.S.A.*, <u>75</u>, 5652
Gantt, E., Lipschultz, C. and Zilinskas, A. (1976)
Biochem. Biophys. Acta, <u>430</u>, 375
Grinvald, A. and Steinberg, I.Z. (1976) *Biochem. Biophys.
Acta,* <u>427</u>, 663
Hindman, J., Kugel,R., Smirmikas, A. and Katz, J. (1978)
Chem. Phys. Letters, <u>53</u>, 179
Hirsch, M.D., Marcus, M., Lewis, A., Mahr, H. and Frigo, N.
(1976) *Biophys. J.* , <u>16</u>, 1399
Hirsch, M.D. and Mahr, H. (1979) *Chem. Phys. Letters,* <u>60</u>, 299
Hughes, R.G. (1975) *Biochem. Essays,*<u>11</u>, 1
Jahnig, F. (1979) *Proc. Natl. Acad. Sci. U.S.A.*, <u>76</u>, 6361
Kawato, S., Kinosita,K. and Ikegami, A. (1977)
Biochemistry , <u>16</u>, 2319
Mahr,H. and Hirsch, M.D. (1975) *Opt. Commun.,*<u>13</u>, 96
Millar, D.P., Shah, R. and Zewail, A.H. (1979) *Chem. Phys.
Letters,* <u>66</u>, 435
Millar, D.P., Robbins, R. and Zewail, A.H. (1980) *Proc. Natl.
Acad. Sci. U.S.A.*, <u>77</u>, 5593
Nicolson, G. (1976) *Biochem. Biophys. Acta,* <u>457</u>, 57
Nigg, E. and Cherry, R.J. (1979) *Nature,* <u>277</u>, 5696
Porter, G., Synowiec, J.A. and Tredwell, C.J. (1977)
Biochem. Biophys. Acta, <u>459</u>, 329
Porter, G., Tredwell, C.J., Searle, G.F.W., and Barber, J.
(1978) *Biochem. Biophys. Acta,* <u>501</u>, 232
Porter, G. and West, M.A. (1978) Chemistry in Microtime
In 'Highlights of British Science', The Royal Society,
London

Searle, G.F.W., Barber, J., Porter, G. and Tredwell, C.J.
    (1978) *Biochem. Biophys. Acta* <u>501</u>, 232
Shank, C.V., Ippen, E.P., Fork, R.L., Migus, A. and
    Kobayashi, T. (1980) *Phil. Trans. Roy. Soc.*<u>298</u>, 93
Swenberg, C., Gaecintov, N. and Pope, M. (1976) *Biophys. J.*
    <u>16</u>, 1447
Welford, D., Sibbett, W. and Taylor, J.R. (1980)
    Opt. Commun.<u>34</u>, 175

STEADY STATE AND DYNAMIC FLUORESCENCE STUDIES OF THE ADAPTATION OF CELLULAR MEMBRANES TO TEMPERATURE

ANDREW R. COSSINS

Department of Zoology
University of Liverpool
Liverpool L69 3BX

INTRODUCTION

It is generally recognised that living organisms are very dynamic and are constantly adjusting their physiology and biochemistry to suit the prevailing environmental conditions, a process commonly termed acclimatization (Prosser, 1973). Such physiological adjustments may be observed in response to a variety of environmental changes such as pressure, light, temperature and salinity and are manifest at the whole animal, organ and cellular grades of organisation. Clearly, acclimatization must have a cellular and molecular basis and the elucidation of the cellular mechanism of response constitutes a major problem in cell biology which is currently being studied in a number of different ways.

The acclimatization of cold-blooded animals to seasonal or diurnal fluctuation in temperature has proved to be a popular subject for study. This is mainly because of the all-pervasive effects of environmental temperature upon such animals, and its obvious ecological importance. Temperature is also a very convenient environmental parameter for study since it is easily and inexpensively measured and may be conveniently adjusted in the laboratory. In addition, there is a voluminous literature describing the effects of temperature variations upon the physiology of animals and its theoretical analysis (Prosser, 1973; Precht *et al.*, 1973; Hazel and Prosser, 1974). Acclimatization of animals to altered temperature may be manifest in two quite distinct ways (Prosser, 1964). Firstly, the rates of an organism's physiological processes may be altered during acclimation in

order to compensate for the direct effects of the altered
temperature. Such compensatory adjustments serve to maintain
the rates of various processes more or less independent of
seasonal variations in temperature. Secondly, the extreme
high and low temperatures that an individual can tolerate may
be adjusted to suit the prevailing conditions more closely.
An example of such resistance is provided in figure 1 where
the ability of goldfish to perform a conditoned avoidance
response at various temperatures was investigated (Heyland
et al., 1979). Goldfish maintained in the laboratory for
several weeks at 10°C (a process termed acclimation in
contrast to acclimatization which refers to adjustments to
the many environmental factors that occur naturally) showed
high success rates over the range 8-27°C, whilst goldfish
acclimated to 30°C were successful between 18-38°C. Clearly,
the temperature range over which fish were behaviourally
competant is adjusted to suit their respective thermal environments.

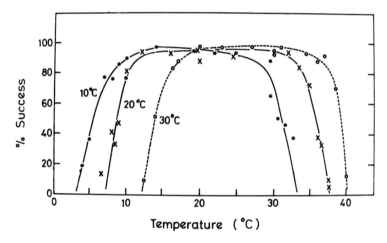

*Fig. 1. The effect of temperature upon the ability of
differently acclimated goldfish to perform a conditioned
avoidance response. Goldfish were acclimated before the
experiment to 10, 20 or 30°C. Heating and cooling scans
were performed separately at least seven days apart (Hoyland
et al., 1979)*

    Recent neurophysiological studies have shown that such
behavioural adaptations are associated with changes in the
performance and temperature sensitivity of their neuronal
machinery (Friedlander *et al.*, 1976). In particular, it
appears that the structures which mediate the transmission
of nervous impulses from one neurone to its neighbour, the

synapse, are particularly sensitive to temperature fluctuations and are critical sites for acclimatization. Indeed,
there is some evidence to implicate adjustments of synaptic
membrane structure during the behavioural adjustments
illustrated in figure 1 (Cossins, 1977; Cossins *et al.*, 1978).
So in this case, at least, it seems possible to suggest a
cellular basis for a whole-animal adaptation to altered
environmental temperature.
     Of course, the integrity and functional properties of
cellular membranes are vital to an enormous number of cellular
processes ranging from the maintenance of ionic gradients to
the activity of membrane-bound enzymes, and all of them will
presumably be affected in some way by temperature changes.
Therefore, it would seem to be highly advantageous for
organisms to maintain membrane function indepedent of seasonal temperature variations so that the strategy of adjusting
membrane structure during temperature acclimation, a process
termed homoviscous adaptation (Sinensky, 1974), may have a
very wide application. Indeed, there have been numerous
reports of altered membrane function and biochemical composition as a result of acclimation treatment (see Hazel and
Prosser, 1974; Cossins, 1976). A seemingly invariable
finding is of increased acyl chain unsaturation at reduced
acclimation temperatures which has usually been interpreted
on the basis of experiments with model membrane systems as
a homeoviscous response. The recent development of spectroscopic techniques for the detection of molecular motion
within the bilayer has permitted the direct demonstration of
homeoviscous responses in *E. coli* (Sinensky, 1974), *Bacillus
stearothermophilus* (Esser and Souza, 1974), *Tetrahymena
pyriformis* (Nozawa *et al.*, 1973) and in various membranes
of teleost fish (Cossins, 1976; Cossins *et al.*, 1978; Cossins
and Prosser, 1978; Cossins *et al.*, 1979; Cossins *et al.*,
1980).
     What follows is an attempt to demonstrate differences in
the dynamic structure of cellular membranes isolated from
fish that have been acclimated to different temperatures
using both steady state and dynamic fluorescence polarisation
techniques, and to elucidate the cellular mechanism of the
adjustment.

STEADY STATE FLUORESCENCE POLARISATION

The hugely popular fluid mosaic model for the structure of
biological membranes lays great emphasis upon the molecular
motion of the constituent molecules (Singer and Nicholson,
1972; Singer, 1974). This motion is manifest in a number of
distinct ways such as the lateral diffusion of lipids and

proteins along the plane of the membrane, the 'flip-flop' of lipids from one monolayer to the other, the flexing of the acyl chains by rotation about their individual carbon-carbon bonds and the wobbling and rotation of the entire molecule about a plane normal to the membrane. Each type of motion occurs over a distinct time course and has a distinctive effect upon the functional properties of the membrane. Nevertheless, the resultant dynamic state is commonly described by the all-inclusive term 'fluidity' and a major problem in membrane biology is the interpretation of such a complex and heterogeneous property in terms of the quantitative spectroscopic techniques now available. For an isotropic bulk liquid, fluidity is simply the reciprocal of viscosity. However, as will be discussed, most membranous environments are not isotropic and exhibit a considerable complexity of organisation and biochemical constitution such that several rotational parameters and rate constants may be required to describe the dynamic nature of a system. Indeed, the expectation of simple relationship between the rates of molecular motion and the degree of molecular order in a system does not apply in a number of instances (Schreier *et al.*, 1978). Thus, it should be borne in mind that any measure of fluidity will refer only to the type of motion sensed by that particular technique and is heavily biased to only one component of the overall molecular motion.

An increasingly popular technique for the estimation of membrane fluidity, and the one used in the studies described here, is the fluorescence polarisation technique. A fluorescent molecule, or fluorophor, is intercalated into the hydrophobic region of the membrane and the rotation that occurs during its fluorescence lifetime is followed by measuring the polarisation of fluorescent light. The technique depends upon the fact that fluorophors tend to absorb incident radiation only if their molecular axis is correctly aligned with respect to the plane of polarisation of the incident light, a process termed photoselection. Similarly, the fluorescent light emitted by an excited fluorophor will tend to be polarised parallel to a particular molecular axis termed the emission vector. In the case of the fluorophor most commonly used in membrane studies, 1,6-diphenylhexatriene (DPH), the emission vector lies parallel to the absorption vector, so that in the absence of molecular rotation the emitted light will be maximally polarised. If, however, molecular rotation occurs during the excited lifetime the emission vector will no longer be parallel to the absorption vector and the emitted light will be depolarised. The degree of depolarisation thus gives a measure of the rate of rotation of the fluorophor about its absorption/

emission vector. If the fluorophors are excited by an infinitely brief flash of light as occurs during time-resolved studies then the polarisation, or equivalently, the anisotropy of the emitted light decays exponentially with time to its minimum value. This is because as time progresses the probes will have rotated farther and farther from their initial orientation before emission of fluorescent light occurs. If, however, the fluorophors are excited by a continuous light source then the value of polarisation or anisotropy will remain constant at a steady-state value that is determined by the integration of the polarisation over the time taken to reach a minimum value (see Shinitzky and Barenholz, 1978).

In the present studies the polarisation of DPH fluorescence was measured using the T-format fluorescence spectrometer described by Jameson *et al*. (1978). The membranous sample was suspended in 0.1 M phosphate buffer solution, pH 7.2 at an absorbance (450 nm) of 0.1-0.2. The probe was introduced by injecting 2 μl of a 2 mM solution in glass-distilled tetrahydrofuran into 2 ml of membrane suspension in a 10 mm quartz fluorescence cuvette and the solution was vigorously stirred and allowed to equilibrate for approximately 30 min at room temperature. The uptake of DPH into the membranes was usually complete within this time during which time the polarisation remained constant. Additional of a further 2 μl aliquot of DPH to the sample did not affect the polarisation value indicating firstly, that the properties of the bulk membrane fluidity was not affected at the concentrations of DPH used and secondly, that excitation transfer between DPH molecules did not occur. The particular advantages of DPH as a fluorescent probe in membrane studies have recently been discussed elsewhere (Shinitzky and Barenholz, 1978).

A schematic diagram of the polarisation fluorescence spectrometer is shown in figure 2. The sample was illuminated by light from a 450 W xenon arc lamp which was selected by a Bausch and Lomb 0.5 m monochromator. The resulting monochromatic radiation (358 nm) was polarised by a double Glan-Foucault calcite prism. Fluorescence from the sample was detected in two channels in which the Glan-Foucault prisms were orientated at right angles to each other. The scattered incident radiation was filtered in each emission channel by a 2mm pathlength of aqueous 2 N sodium nitrite solution followed by a Corning CS 3-73 sharp cut-off filter. Scattered radiation typically comprised less than 1% of the total detected radiation and was corrected using an identical sample but without added probe as described by Shinitzky *et al*. (1971). In later experiments, a 4 mm Corning 7-54 broad

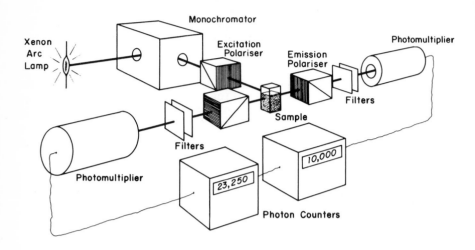

Fig. 2. A schematic diagram of the T-format polarisation fluorescence spectrometer used for the measurement of steady state fluorescence polarisation (Jameson et al., 1978)

band-pass filter was inserted into the incident beam to reduce transmission of higher order wavelengths. Scattered radiation was reduced to less than 0.5% total detected radiation and the correction was not necessary.

Polarisation of fluorescence (p) and anisotropy (r) is usually calculated from :

$$p = \frac{I_{\parallel} - I_{\perp}}{I_{\parallel} + I_{\perp}} \qquad (1)$$

$$r = \frac{I_{\parallel} - I_{\perp}}{I_{\parallel} + 2I_{\perp}} \qquad (2)$$

where $I_{\parallel}$ and $I_{\perp}$ are the intensities of fluorescent light detected through polarisers orientated parallel and perpendicular, respectively, to the excitation light. In the present experiments, the fluorescence was detected by photon counters which provide a digital output. The circuitry was arranged to stop counting when the $I_{\parallel}$ value reached 10^6 so that the $I_{\parallel}$ value represented the ratio of $I_{\parallel}$ to $I_{\perp}$. This was inserted into the following equations :

$$p = \frac{(I_{\parallel}/I_{\parallel}) - 1}{(I_{\parallel}/I_{\parallel}) + 1} \qquad (3)$$

$$r = \frac{(I_{\parallel}/I_{\perp}) - 1}{(I_{\parallel}/I_{\perp}) + 2} \qquad (4)$$

The equation which related fluorescence polarisation to the rotational characteristics of the probe is known as the Perrin equation (Perrin, 1926). In the case of a spherical fluorophor which rotates isotropically in a bulk solvent, the rate of rotation, $\bar{R}$ (radians s^{-1}) or alternatively the rotational relaxation time (σ) is obtained from :-

$$\frac{r}{r_0} = \frac{(1/p) - (1/3)}{(1/p_0) - (1/3)}$$

$$\simeq 1 + 6\bar{R}\tau = 1 + \frac{3\tau}{\sigma} \qquad (5)$$

where p_0 and r_0 are the upper limits of polarisation and anisotropy, respectively, measured in the absence of molecular rotations and for this instrument was 0.485 (Fraley *et al.*, 1978); τ is the fluorescence lifetime which is determined as described presently. It is worth noting at this point that the Perrin equation assumes that the rotational motion of the fluorophor is unhindered and isotropic, that the population of fluorophors is homogeneous with respect to rotational motion and that the decay of fluorescence following excitation with a pulse of light is monoexponential.

FLUORESCENCE LIFETIME DETERMINATIONS

The fluorescence lifetime was measured using phase fluorometry (Spencer and Weber, 1969) in which the intensity of the monochromatic excitation light was modulated sinuosidally at a frequency of 18 MHz by an ultrasonic modulator (figure 3). The fluorescence from the sample that was detected at right angles to the excitation beam was also modulated sinusoidally at the same frequency but was slightly out of phase with respect to the excitation beam because of the finite time delay between absorption of incident light and emission of fluorescent light (figure 4). The tangent of the phase angle (tan δ) was related to the fluorescence lifetime, τ, by

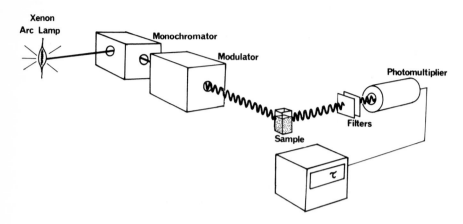

Fig. 3. Schematic diagram of the phase fluorometer used for the measurement of fluorescence lifetime. The excitation beam was modulated (wavy line) by the Sears-Debye ultrasonic modulator at 18 MHz (Spencer and Weber, 1969)

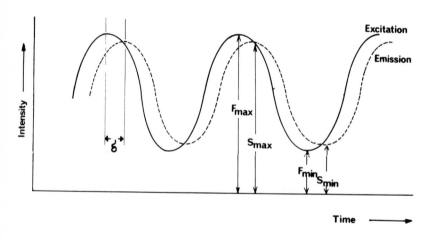

$$D_f = F_{max} - F_{min} \Big/ F_{max} - F_{min}$$

$$D_s = S_{max} - S_{min} \Big/ S_{max} - S_{min}$$

Fig. 4. Schematic diagram showing the phase lag and demodulation of the photocurrents with a scatterer (such as glycogen) in the sample cuvette (excitation) and with a fluorophor in the sample cuvette (emission). D_f and D_s refer to the modulation of fluorescence and excitation, respectively (Spencer and Weber, 1969)

$$\tan \delta = 2 \pi f \tau \qquad (6)$$

where f is the modulation frequency (Spencer and Weber, 1969). In addition, the emitted light was demodulated with respect to the excitation beam (figure 4) so that :

$$\text{Relative modulation} \equiv D = \frac{\text{Modulation of fluorescence}}{\text{Modulation of excitation}}$$

$$= \cos \delta = (1 + 4\pi^2 f^2 \tau^2)^{\frac{1}{2}} \qquad (7)$$

Thus it was possible to determine fluorescence lifetimes either by measurement of phase lag or by relative modulation. For a homogeneous population of fluorophors the lifetime determined by both methods should be identical. In a heterogeneous emitting population, however, the lifetime determined by the modulation method is usually longer than the true weighted average of the component lifetimes, whereas the lifetime determined by phase lag will be shorter than the weighted average. The experimental observation of differences in the lifetime determined by the two methods can, therefore, serve as a semi-quantitative index of the heterogeneity of the emitting population (Spencer and Weber, 1969). The phase method has an important advantage over conventional nano-second pulse techniques for the study of labile biological preparations since each measurement can be performed in 1-2 min and no further data analysis is necessary.

SEASONAL HOMEOVISCOUS ADAPTATION

The effects of temperature on the fluidity of brain synapto-somal membranes in goldfish are illustrated in figure 5 (a-c). An increase in temperature invariably resulted in a dramatic increase in $\overline{R}$ (figure 5c) and, by inference, in the degree of membrane fluidity. Of prime interest in the present context was the different positions occupied by the curves for membranes isolated from 5°C and 25°C acclimated goldfish. The graph for 5°C acclimated fish was shifted to lower temperatures compared to the corresponding graph for the 25°C acclimated fish. This indicates that the membranes at 5°C acclimated fish were more fluid than those of 25°C acclimated fish at all measurement temperatures. This increase partially compensates for the rigidifying effects of the temperature reduction and thus maintains fluidity at least partially independent of seasonal fluctuations in temperature. The values of $\overline{R}$ were highly reproducible (table 1) indicating

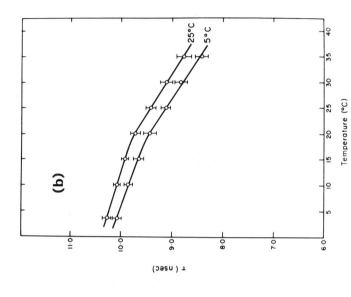

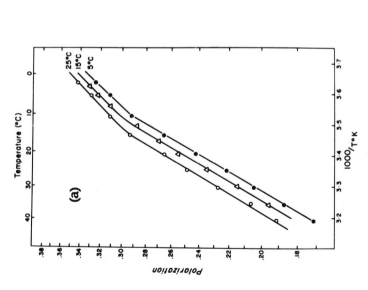

Fig. 5. The effects of temperature upon the polarisation (a) and fluorescence lifetime (b) of DPH in brain synaptosomes of goldfish acclimated at 5, 15 and 25°C. The fluorescence lifetime was calculated as the average of values obtained by phase and modulation methods (see text). Values represent the mean ± standard error of the mean (Cossins, 1977)

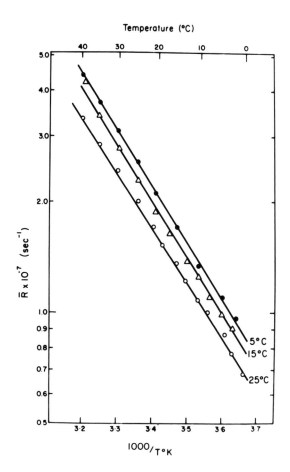

Fig. 5(c). The effects of thermal acclimation of goldfish upon the rotational diffusion coefficient (R̄) of DPH in brain synaptosomes. Goldfish were acclimated to 5, 15 and 25°C (Cossins, 1977)

not only that the technique was capable of great precision but also that goldfish are able to regulate membrane fluidity to a fine degree. The slopes of the Arrhenius plots were not affected by acclimation treatment. It should be noted that both polarisation and fluorescence lifetime (figures 5a and 5b) have a curvilinear or perhaps biphasic relationship with temperature which may be interpreted as evidence of a thermotropic restructuring of the probe environment such as a phase transition or separation. However the Arrhenius plot of the rotational diffusion coefficient, R̄, was precisely linear over the temperature range 3-40°C and such a conclusion

TABLE 1
*The average rotational diffusion coefficient ($\overline{R}$) at 5,
15 and 25°C for DPH in synaptosomal membranes of 5, 15
and 25°C acclimated goldfish*

	$\overline{R}$ x 10^7 (s^{-1})			
	5°C	15°C	25°C	E_a ** (kcal.mole^{-1})
5°C acclimated goldfish (n = 4)	1.08* ±0.02	1.67 ±0.02	2.55 ±0.02	6.87 ± 0.13
15°C acclimated goldfish (n = 2)	0.96	1.48	2.25	7.18
25°C acclimated goldfish (n = 4)	0.86 ±0.02	1.32 ±0.03	1.98 ±0.04	6.95 ± 0.11

* Values represent the mean ± S.E.M. of n measurements
**E_a represents the Arrhenius activation energy

was clearly inappropriate. There was no reason to expect a
linear Arrhenius plot of polarisation for an unchanging
system since polarisation was not a rate constant and therefore does not necessarily conform to the Arrhenius formulation.

In figures 6 (a-d), the differences observed in several
different membranous preparations of variously acclimated
fish has been illustrated as a graph of $\overline{R}$ measured at 15°C
plotted against acclimation temperature. Compensation was
apparent in the goldfish brain synaptosomes (Cossins, 1977),
figure 6a and green sunfish liver mitochondria (Cossins *et
al.*, 1980), figure 6d as a negative slope since $\overline{R}$ (at 15°C),
or fluidity, was greater for animals acclimated at lower
temperatures. By contrast, the fluidity of goldfish sarcoplasmic reticulum was not significantly affected by thermal
acclimation (Cossins *et al.*, 1978) although in this membranous system the variability between preparations was quite
large. Liver microsomes of green sunfish exhibited much
reduced compensatory responses between 5 and 25°C whilst the
34°C acclimated fish were significantly less fluid than the
25°C acclimated fish (Cossins *et al.*, 1980).

Clearly, different subcellular membrane preparations even
within the same cell exhibited quite different homeoviscous
responses, but in no case has the compensation observed by
steady state polarisation techniques been completely effective

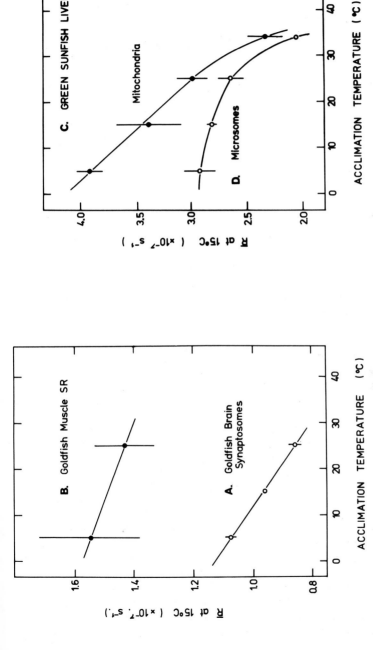

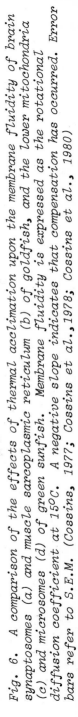

Fig. 6. A comparison of the effects of thermal acclimation upon the membrane fluidity of brain synaptosomes (a) and muscle sarcoplasmic reticulum (b) of goldfish, and the lower mitochondria (c) and microsomes (d) of green sunfish. Membrane fluidity is expressed as the rotational diffusion coefficient at 15°C. A negative slope indicates that compensation has occurred. Error bars refer to S.E.M. (Cossins, 1977; Cossins et al., 1978; Cossins et al., 1980)

such that membrane fluidity was identical at all acclimation
temperatures. However, this does not necessarily imply that
the homeoviscous response was incomplete in a functional
sense. Firstly, adaptations of other membranous components
such as membrane-bound proteins may contribute to the overall
adapted state (Sidell, 1977). Secondly, the optimal condition
towards which the system was becoming adjusted may itself vary
with temperature. Thirdly, there were certain limitations both
to the steady state polarisation technique and to the experimental approach adopted here which affect the conclusions
drawn heretofore. The distribution of the probe within the
membrane and between various membranes of each preparation
was not known with any certainty. DPH was certainly located
within a hydrophobic environment since it does not fluoresce
in aqueous media (Shinitzky and Barenholz, 1974, 1978). The
available evidence indicates that DPH does not partition
preferentially either into the gel or fluid phases of a binary
membrane system (Andrich and Vanderkooi, 1976; Lentz et al.,
1976) so that the fluidity estimates provided by steady state
techniques are the weighted average for the many different
environments within each membrane-type. For this reason,
precise correlations of the estimated fluidity with specific
membrane functions, which may be associated with specific
microdomains, will be difficult to achieve. It is a central
assumption of the present study that the preparations isolated from warm and cold acclimated fish have similar distributions of DPH binding sites and that these are not significantly affected by temperature.

     A problem common to all techniques which utilise exogenous
spectroscopic probes is the degree of disturbance to both the
bulk membrane fluidity and the immediate environment of the
probe itself, since this will be the environment sensed by
the probe and upon which the estimates of fluidity are based.
Several studies have demonstrated that at the membrane
concentrations of probe that are commonly used in fluorescence
polarisation studies, the bulk properties of chemically-
defined, artificial membranes are not significantly perturbed
(Jacobson and Papahadjopoulos, 1975; Andrich and Vanderkooi,
1976; Suurkuusk et al., 1976). On the other hand, studies
with non-perturbing techniques such as NMR have shown some
perturbation of the immediate environment of bulky ESR probes
(Seelig, 1977). Cadenhead et al. (1975) have demonstrated
with monolayer experiments that doxyl-labelled fatty acids
cause a greater perturbation than the analogous fluorescence
probe. However, whatever perturbation of membrane structure
may be caused by DPH, it is reasonable to expect that it will
apply to an equal degree in the membranes of both cold and
warm-acclimated fish. Thus, although the estimated fluidity

may not have any absolute basis, the differences observed
between preparations will accurately reflect differences in
membrane fluidity.

THE MECHANISM OF HOMEOVISCOUS ADAPTATION

The alteration in membrane fluidity that accompanies thermal
acclimation is invariably associated with changes in the biochemical composition of the membrane protein and lipids
(Hazel   and Prosser, 1974). As a first step to identifying
the nature of the compositional adjustments, liposomes were
prepared from the purified phospholipids of synaptosomal
membranes at 5 and 25°C acclimated goldfish and their fluidity
was compared with steady state polarisation techniques
(Cossins, 1977), figure 7. The rotational diffusion coefficient in the pure phospholipid membranes was approximately
twice that of the corresponding natural synaptosomal membranes,
indicating a much more restrictive probe environment in the
latter that may be associated with the presence of rigidifying elements such as intrinsic membrane-bound proteins and
cholesterol. Nevertheless, the liposomes prepared from cold-
acclimated fish were more fluid that those of warm-acclimated
fish. Although membrane fluidity may be influenced by all
membranous constituents, it is clear that differences in
synaptosomal membranes of cold- and warm-acclimated fish were
explained in large part by changes in the biochemical composition of their constituent phosphatides.
The fatty acid composition of the major phospholipid classes
have been studied in detail and table 2 shows the results of
a typical analysis for choline phosphoglycerides of the liver
microsomes of 5°C and 25°C acclimated green sunfish. The
mixture of fatty acids was very complex with chain lengths
varying between 16 and 22 carbon atoms and with up to six
olefinic bonds. There were substantially smaller proportions
of saturated fatty acids and correspondingly greater proportions of unsaturated fatty acids at the lower acclimation
temperatures, which was consistent with the large number of
previous compositional studies on micro-organisms, plants and
animals (see for example Caldwell and Vernberg, 1970; Cullen
et al., 1971; Aibara et al., 1971; Fukushima et al.,1976).
Olefinic bonds introduce a slight misalignment into the acyl
chain which results in a greater steric hindrance to the
close approach and packing of adjacent acyl chains (Salem,
1962). Hence, the more olefinic bonds that are present in
the hydrophobic core of the membrane the greater is the cross-
sectional area and swept volume occupied by each phosphoglyceride and the greater is the potential for molecular
motion. Beyond this generalisation, our understanding of the

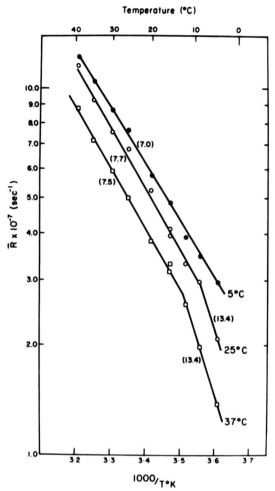

Fig. 7. The effects of thermal acclimation of goldfish upon the membrane fluidity ($\bar{R}$) of liposomes prepared from the purified phospholipids of goldfish brain synaptosomes. For comparison, the corresponding data for liposomes prepared from rat synaptosomes have been included. The values in brackets refer to the Arrhenius activation energies (Kcal. mole^{-1}) for the adjacent graph

structure/fluidity relationship of acyl chains is quite rudimentary. For example, there is no obvious functional reason for the occurrence of enormous quantities of docosahexaenoic acid ($22:6\omega3$) in the retinal rod membranes of mammals, or the occurrence of $22:5\omega3$ as the major polyene in decapod crustacea (Cossins, 1976) and $22:6\omega3$ in teleost fish

TABLE 2

The fatty acid composition of the choline phosphoglyceride fraction isolated from liver mitochondrial fraction at 5°C and 25°C acclimated green sunfish

Fatty Acid	5°C Green Sunfish (n=3)	25°C Green Sunfish (n=4)
16:0	13.6 ± 1.3*	18.1 ± 0.5
16:1	6.1 ± 0.9	4.3 ± 0.1
18:0	4.7 ± 0.4	5.7 ± 0.2
18:1	16.9 ± 0.5	11.3 ± 0.6
18:2ω6	6.6 ± 0.2	14.4 ± 0.9
18:3ω3	2.7 ± 0.3	2.6 ± 0.2
20:4ω6	0.9 ± 0.3	2.8 ± 0.3
20:5ω3	0.7 ± 0.1	1.6 ± 0.1
22:5ω3	2.2 ± 0.1	1.5 ± 0.1
22:6ω3	41.9 ± 1.5	30.8 ± 0.5
Others**	3.8	6.9
Total saturated	19.2 ± 1.7	25.6 ± 0.4
Total Monounsaturated	23.4 ± 1.0	16.4 ± 0.7
Total Polyunsaturated	57.2 ± 2.0	58.0 ± 0.4
Unknown	-	-
Unsaturation Index***	320.2 ± 10.4	278.1 ± 3.3

*   Values represent the mean ± S.E.M. for n separate preparations
**   Other components present in trace quantities include 17:1ω9, 20:0, 20:2ω6, 22:0, 22:2ω6, 22:4ω6 and 22:5ω6
***   Unsaturation index was calculated as the sum of the % weight multiplied by the number of olefinic bonds for each fatty acid in the mixture.

(Cossins, 1977). Certainly, fluidity will not depend simply
upon the number of olefinic bonds in a mixture of phospholipids since the position of the bonds on the carbon chain,
the number on each chain and, perhaps, the specificity of
acyl incorporation into the 1 and 2 position of phospholipids
may all have some influence on their membrane properties
(Miller et al., 1976).
   The gross differences in acyl group composition of 5^0C and
25^0C acclimated fish are illustrated in figure 8. For brain
synaptosomes, (figure 8a) the differences were as expected,
with substantially higher proportions of saturated fatty
acids in all phosphoglyceride fractions of 25^0C acclimated
goldfish, and with corresponding reductions in the proportion
of both monounsaturated and polyunsaturated fatty acids. The
sarcoplasmic reticulum of goldfish (figure 8b) also displayed
large differences in acyl group composition, but in this case
the proportion of saturated fatty acids was only slightly
higher in 25^0C acclimated fish particularly in the choline
phosphoglycerides. Contrary to the general trend there were
substantially higher proportions of monoenes in 25^0C
acclimated fish compared to 5^0C acclimated fish. It is known
that the incorporation of an additional olefinic bond into
an already unsaturated acyl chain certainly causes a less
dramatic shift in the physical properties of the fatty acid
than does the unsaturation of a saturated fatty acid (Ghosh,
1971). We might, therefore, expect that the exchange of
monounsaturated for polyunsaturated as occurs in the sarcoplasmic reticulum may have no measurable effect upon membrane
fluidity which is in agreement with the lack of homeoviscous
response in this preparation. Interestingly, a quite different
mechanism for the adaptation of sarcoplasmic reticulum function
has recently been suggested (Penny and Goldspink, 1980),
namely the elaboration of a more intimate reticulum at
reduced acclimation temperatures.
   A similar distinction between the compositional differences
of membranes which exhibit good and poor homeoviscous responses
are shown in figure 8 (c-d) for the liver microsomes and
mitochondria of green sunfish; a very eurythermal fish that
is indigenous to the central United States. In mitochondria,
which show marked homeoviscous responses, the increased
proportion of saturated fatty acids in warm acclimated fish
is quite dramatic. In the choline phosphoglycerides of
microsomes, which comprise over 50% of the membrane phosphoglycerides, there was only a moderate difference in the
saturated fatty acids but a comparatively large exchange
between the monoenes and polyenes.
   This compositional data leads to two main conclusions.
Firstly, the details and magnitude of the compositional

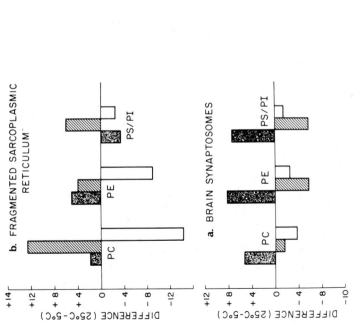

Fig. 8. Histograms to illustrate the difference in acyl group composition of various membrane preparations of 5°C and 25°C acclimated goldfish (a,b) and green sunfish (c,d). The difference is expressed as the average % weight for the unsaturated, monounsaturated and polyunsaturated fatty acids for the 25°C acclimated fish minus the corresponding value for the 5°C acclimated fish. PC— choline phosphoglycerides, PE— ethanolamine phosphoglycerides, PS/PI — serine— inositol phosphoglycerides.

adjustment was complex but based on certain assumptions can
be correlated well with the magnitude of the homeoviscous
response observed by steady state polarisation methods.
Secondly, there are very significant alterations in the lipid
composition of some membranes with no measurable effects on
fluidity, so that the observation on altered fatty acid composition does not necessarily mean that membrane fluidity is
significantly affected. Evidence of a structural or functional adaptation must be obtained separately.

TIME COURSE OF HOMEOVISCOUS ADAPTATION

The period of time necessary for homeoviscous adjustments has
been studied in goldfish synaptosomes both for acclimation to
cold (i.e. transfer from 25 to 5^{0}C) and for acclimation to
warm (i.e. from 5 to 25^{0}C) (Cossins *et al.*, 1978). Groups
of approximately 40 goldfish were transferred from 15^{0}C to
constant temperature rooms at 5^{0}C or at 25^{0}C where they were
left for at least 3 weeks to become fully acclimated to their
respective temperatures. Each group was then transferred to
the alternate temperature and at various intervals, groups of
six fish were killed and the fluidity of their synaptosomal
membranes were estimated by measuring the fluorescence
polarisation at 25^{0}C.
   During warm acclimation (figure 9a), membrane fluidity
increased rapidly until after approximately 14 days it was
identical to the value for fish that had been acclimated for
a long time at 25^{0}C. By contrast, transfer from 25 to 5^{0}C
(figure 9b) resulted in no significant change in fluidity
for 20 days after which there was a rapid change to the value
characteristic of 5^{0}C acclimated fish. The time taken for
homeoviscous adjustments were somewhat larger than expected
but compare favourably with the time-course of the behavioural
adjustment of the whole animal (Cossins *et al.*, 1977). The
apparent lag during the cold-transfer experiment presumably
reflects a phase during which the mechanism of adjustment was
affected by the dramatic and rapid fall in temperature.
Analysis of the acyl group composition of synaptosomal
membranes isolated during the transfer experiment indicated
a good correlation between the adjustment of fluidity and the
ratio of saturated to unsaturated fatty acids, but not to an
index of the number of olefinic bonds in the mixture (Cossins
et al., 1978).

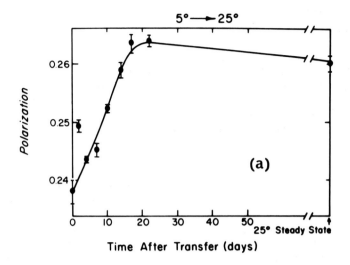

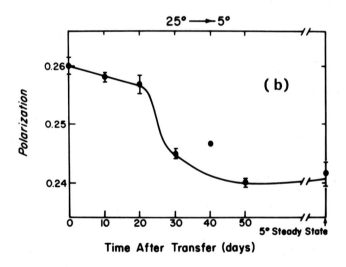

Fig. 9. The time-course of changes in the membrane fludity of brain synaptosomes of goldfish during reacclimation from (a) 5⁰ to 25⁰C and from (b) 25⁰ to 5⁰C. The index of fluidity in this experiment was the value of the fluorescence polarisation at 25⁰C. 'Steady state' values of fluidity were obtained for fish that were acclimated for at least six weeks to their respective temperatures.

EVOLUTIONARY ADAPTATION OF MEMBRANE FLUIDITY

Goldfish and green sunfish inhabit environments which experience dramatic seasonal fluctuations in temperature, and we
have seen how such eurythermal fish are able to adjust the
structure of some cellular membranes to maintain them at
least partially independent of seasonal temperature changes.
Other fish species and mammals live in relatively extreme
and unvarying thermal environments. For example, the fish
family *Nototheniidae* live exclusively in the Antarctic
Ocean at approximately -1°C throughout the year, whilst
mammals and birds have a core temperature that deviates little
from 37-40°C. One might expect that the selection pressure
for cellular adaptation to such extreme thermal environments
to be particularly strong. This together with the unvarying
nature of such cellular environments suggest that the
potential for seasonal flexibility exhibited by eurythermal
species such as goldfish need not be maintained, and the
organism can adapt over evolutionary time more completely to
its particular environment. We have examined this hypothesis
by comparing the fluidity of the synaptosomal membranes of
eurythermal species, such as goldfish and the green sunfish,
with two relatively stenothermal fish species (Cossins and
Prossor, 1978). The arctic sculpin,*Myoxocephalus verrucosus*
was collected in the Bering Sea by Dr. A.L. DeVries at
approximately -0.3°C and was maintained in the laboratory at
0 $\pm$ 1°C. The desert pupfish, *Cyprinodon nevadenis*, were
reared at 28°C by Dr. S.D. Gerking, University of Arizona,
from stocks originally obtained from Death Valley, California.
They were warmed over a 4 day period to 34°C where they were
maintained for 7 days before the experiment.
    Arrhenius plots of polarisation for the various preparations
are presented in figure 10. The experiments on the arctic
sculpin and the desert pupfish were carried out simultaneously
with analyses of 5°C and 25°C acclimated goldfish, respectively, in an effort to control for the minor differences of
preparative technique upon the comparison. However, the
results for two 5°C acclimated goldfish and two arctic sculpin
preparations performed 7 days apart were identical, demonst-
rating again the good reproducibility of the observations.
For reasons given previously no significance is attached to
the non-linear graphs. Figure 10 clearly shows a shift of
curves upwards and to the left with higher acclimation,
habitat or body temperature. Thus, over the entire temperature range, the synaptosomal membranes of the arctic sculpin
were considerably less restrictive and hence more fluid than
those of the 5°C acclimated goldfish, whilst the membranes
of the desert pupfish had a fluidity between those of the

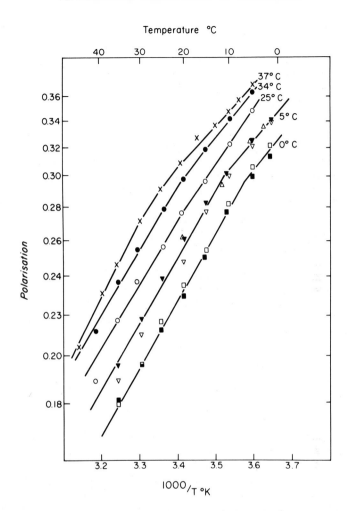

Fig. 10. A comparison of membrane fluidity between 3°C and 42°C of brain synaptosomes isolated from the arctic sculpin (Myoxocephalus sp., 0°C,☐■), 5°C acclimated goldfish (∇ ▼), 25°C acclimated goldfish (o), the desert pupfish (34°C, ●) and the rat (37°C, x). Each symbol represents a separate preparation.

25°C acclimated goldfish and the rat.

These results are summarised in figure 11 as a graph of polarisation for each preparation at its respective cellular temperature plotted against cellular temperature. Complete or perfect compensation would be apparent as a horizontal line, as can be seen for the comparison of the arctic sculpin with 5°C acclimated goldfish, and the comparison of the desert

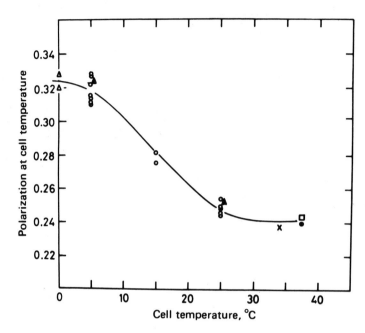

Fig. 11. The efficacy of homeoviscous compensation over the seasonal and evolutionary time-scales illustrated by a graph of membrane fluidity at the cell temperature of each preparation plotted against cell temperature. If compensation was complete then the graph would have zero slope. Arctic sculpin (Δ), goldfish (o), green sunfish (▲), desert pupfish (x), rat (●) and hamster (□). Each point represents an individual animal.

pupfish and the rate with the 25°C acclimated goldfish. The eurythermal species, the goldfish and green sunfish, exhibited incomplete compensation when acclimated to 5°C, 15°C and to 25°C since the values at each acclimation temperature were not identical. It appears, therefore, that species that have evolved in constant and extreme thermal habitats can permit a more complete specialisation of their membranes to their respective cellular temperatures and that this may be associated with the loss and their ability to acclimate over a seasonal time-scale. The fact that the membranes of rat and hamster fall into the general sequence suggest a more general relationship between cellular temperature and synaptosomal membrane fluidity, which crosses broad phylogenetic boundaries.

The fatty acid composition of the various membranous preparations was also determined and the results are summarised in table 3 as two indices, the ratio of saturated to

TABLE 3

Summary of the fatty acid composition of brain synaptosomes isolated from various fish species and the rat

Phospho-glyceride		Arctic sculpin		Goldfish		Desert pupfish	Rat
		r	0°C	5°C	25°C	34°C	37°C
Ratio saturated to unsaturated fatty acids	Choline	0.970	0.593	0.659	0.817	0.990	1.218
	Ethanolamine	0.947	0.260	0.340	0.506	0.568	0.651
	Serine/inositol	0.811	0.477	0.459	0.633	0.616	0.664
Unsaturation index *	Choline	0.938	220.8	172.5	149.5	102.8	98.5
	Ethanolamine	0.830	333.1	317.2	316.4	227.6	250.5
	Serine/inositol	0.558	254.8	294.3	284.8	177.0	233.4

r, Correlation coefficient for the dependence of membrane fluidity as expressed by polarisation at 25C upon the various indices of fatty acid composition for each phosphoglyceride class.

*Calculated as sum of weight% multiplied by number of olefinic bonds for each fatty acid in the mixture.

unsaturated fatty acids and as an unsaturation index which
gives a relative estimate of the number of olefinic bonds in
a mixture of fatty acids. There was an uninterrupted trend
towards increased fatty acid unsaturation with lower cellular
temperature. In general, the correlation between membrane
fluidity, as expressed by the polarisation values, and the
saturation ratios was better than between membrane fluidity
and unsaturation index (figure 12). The difference in membrane
fluidity of the various species appears to be principally
determined by differences in the acyl group composition of
their membrane phosphoglycerides and both are related to
their respective cellular temperature despite differences in
dietary lipids, idiosyncracies of lipid metabolism and other
structural modifications that may have arisen during
evolutionary development.

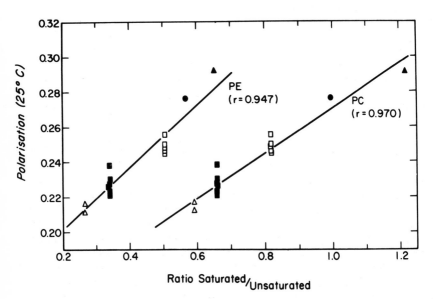

*Fig. 12. The correlation of membrane fluidity of synaptosomal
membranes of various fish species and rat with the acyl group
composition of choline (PC) and ethanoloamine (PE) phosphoglycerides. Each data point represents an individual animal.
Fluidity is expressed as the steady state polarisation of
DPH at 25ºC and the lipid composition is expressed as the
ratio of saturated to unsaturated fatty acids. Symbols :
arctic sculpin (Δ), 5ºC acclimated goldfish (■), 25ºC
acclimated goldfish (□), desert pupfish (●), rat (▲).
r represents the correlation coefficient for each group.*

DYNAMIC STUDIES OF MEMBRANE STRUCTURE

The interpretation of the fluorescence polarisation of
membrane-bound probes depends critically upon the validity
of a number of assumptions applied in the derivation of the
Perrin equation. These are that the depolarising rotations
of the probe are isotropic and unhindered, that the decay of
emission and emission anisotropy following a flash of light
is mono-exponential, and finally, that the population of
fluorophores is homogeneous with respect to their membranous
environment and rotational characteristics. Recent studies
using time-resolved anisotropy measurements have indicated
that the decay of emission anisotropy following a pulse of
light is complex (Chen et al., 1977; Dale et al., 1977;
Kawato et al., 1977; Hildenbrand and Nicolau, 1979). For
example, in sonicated and multilamellar liposomes and cellular
membranes, the decay of anisotropy was clearly resolved into
an initial rapid component followed by a second almost
constant phase. These observations indicated considerable
restriction to the free rotational motion of the probe and
thus have been interpreted using a model (Kinosita et al.,
1977) in which the probe underwent rapid but restricted
rotation within a cone that was defined by the anisotropic
fatty acid environment. The plateau value of emission
anisotropy, r_∞ , reached at times that were long compared to
the fluorescence lifetime, was thought to reflect the degree
of constraint imposed upon the wobbling motion of the probe;
the greater the value of r_∞ , the greater were the restrictions
to probe rotation. Two further assumptions of the model were
that the rate of motion of the probe throughout the cone was
constant and that the orientational distribution of DPH
within the probe at equilibrium was uniform. Interestingly,
the rotational rate shows an exponential increase with temperature, whilst r_∞ has a sigmoidal dependence with a dramatic
decrease at the phase transition temperature (Kawato et al.,
1977). The dramatic changes of steady state anisotropy at the
phase transition temperature, therefore, appears to be
determined principally by the changes to the hindrance to
free rotations (i.e. r_∞) and not by the rotational rate, R
(Lakowicz, Prendergast and Hogan, 1979; Hildenbrand and
Nicolau, 1979).
    Even for an artificial membrane, the Perrin equation is a
considerable oversimplification and the unequivocal description of membrane fluidity with its use is not possible. The
values of rotational diffusion coefficient, $\bar{R}$, rotational
correlation times and microviscosity that are derived from
the Perrin equation contain contributions from both R and r_∞
of the restricted rotational model and are, thus, considerably

less well defined parameters. The concept of microviscosity
was also shown to be somewhat less than rigorous, since the
rotational characteristics of DPH in the membrane and the
reference solvent differ quantitatively and qualitatively
(Chen *et al.*, 1977; Dale *et al.*, 1977), and the absolute
values obtained depend to some extent on the reference solvent
chosen (Hare and Lussan, 1977). Recently, Hare and Lussan
(1978) have proposed a more elaborate scheme for the calibration of microviscosities which involves the use of reference solvents.

   The assumed monoexponential decay of fluorescence intensity
and the homogeneity of probe molecules is also inadequate.
Chen *et al.*, (1977) and Dale *et al.*, (1977) have shown that
even in the case of a chemically-defined, artifical membrane
composed of only one phosphoglyceride species, the decay of
DPH emission is best interpreted in terms of a double
exponential decay and the situation is, presumably, more
complex in natural membranes. The reasons for this complex
decay are not enirely clear although it seems likely that it
may be accounted for by the possible heterogeneity of probe
sites or perhaps by reversible excited-state reactions of
the fluorophor (Chen *et al.*, 1977; Dale *et al.*, 1977; Cehelnik
et al., 1975; Mason and Cehelnik, 1978). However, the
majority component (78-87%) of the double exponential decay
has a lifetime which differs only slightly (0.6 ns) from the
weighted average of both components (Chen *et al.*, 1977), a
difference which probably does not significantly affect the
conclusions drawn from steady state studies. This is confirmed
by harmonic studies which show that the lifetimes measured
by the phase shift and modulation methods, which more heavily
weight the shortest and longest lifetimes, respectively, in
a population of fluorophors, are different, although only by
a small amount (< 0.6 ns) (Cossins, 1977; Cossins *et al.*,
1978).

   Recently, the results of time-resolved anisotropy studies
of membrane structure have been confirmed using a completely
different technique, namely, differential polarised phase
fluorometry (Lakowicz and Prendergast, 1978a,b; Lakowicz *et
al.*, 1979). The theoretical basis for such measurements was
developed by Weber (1977) and was subsequently used to demonstrate the occurrence of anisotropic rotations by aromatic
fluorophors. Weber (1978) has recently extended the theory
to include cases of hindered rotations and Lakowicz and
Prendergast (1978a) have described a method for the calculation
of the limiting anisotropy, r_∞, and the rotational rate, R,
by the combination of data from steady state and differential
polarised phase measurements. We have recently applied the
differential phase technique to the comparison of membranes

isolated from thermally-acclimated fish, in order to provide
a more rigorous and detailed description of the structural
modifications induced by homeoviscous adaptation (Cossins,
Kent and Prosser, 1980).

DIFFERENTIAL POLARISED PHASE FLUOROMETRY

Essentially, the technique requires the measurement of the
difference in fluorescence lifetimes of the parallel and perpendicular components of the emission, the so-called differential lifetime ($\Delta\tau$), the emission that is observed through
polarisers orientated parallel and perpendicular to the plane
of polarisation of the excitation light decay at different
rates. This is because the parallel polariser tends to
select those fluorophors which have emitted their photon
before much rotation has occurred, whereas the perpendicular
polariser tends to select those fluorophores which have
rotated significantly during their excited lifetime. In
harmonic systems, where the fluorophor is excited by a
modulated beam of monochromatic light, this difference in
lifetime is manifest as a differential delay of the modulated
fluorescence in each channel with respect to the modulated
excitation; that is, the phase delay of the parallel component
with respect to the excitation is less than that of the
perpendicular component. This difference in phase delays or
phase angles is referred to as the differential tangent
(tan Δ) which is related to the differential lifetime ($\Delta\tau$) by

$$\tan \Delta = 2 \Pi f \Delta\tau \qquad (8)$$

where f is the modulation frequency.
    Measurements of $\Delta\tau$ for DPH in cellular membranes were
conveniently and rapidly performed on an instrument used for
the fluorescence lifetime measurements except that the
excitation and emission channels were polarised with calcite
prism polarisers (figure 13). Modulation frequency was 18
MHz in all experiments. Filters were placed in excitation
and emission beams as described previously to reduce the
contribution of scattered light to the total detected light.
To measure the differential phase between the polarised
components of the emission, one of the two polarisers was
rotated so that it was parallel to the other polariser and
the phase difference was nulled. One of the polarisers was
subsequently rotated to the perpendicular position and the
phase difference between channels was measured. For each
measurement the average of 100 determinations was presented,
thereby improving the signal to noise ratio by tenfold. This
procedure was repeated until a satisfactory average with a

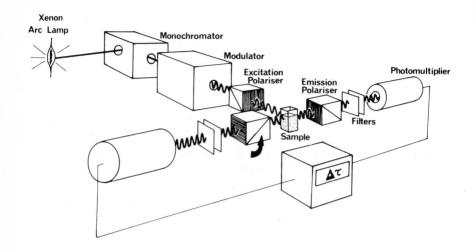

Fig. 13.   A schematic diagram of the phase fluorometer used for the differential polarised phase fluorometric analysis of membrane fluidity

standard deviation of less than 0.05 ns was obtained.

For an isotropic, unhindered fluorophor, the tangent of the differential phase angle (tan Δ) between the parallel and perpendicular components of the emission is given by (Weber, 1978) :

$$\tan \Delta = \frac{(2\ R\tau)\ r_0\omega\tau}{1/9\ m_0\ (1 + \omega^2\tau^2) + 1/3(2R\tau)(2 + r_0) + (2R\tau)^2} \quad (9)$$

where $m_0 = (1 + 2r_0)(1 - r_0)$ \hfill (10)

Knowing tanΔ the calculated value of the rotational rate R can be obtained from the quadratic form of equation 1. For an isotropic unhindered rotator this should agree with the rotational diffusion coefficient ($\overline{R}$)obtained by steady state methods. The maximal value of tanΔ may be calculated from (Weber, 1979) :

$$\tan\Delta_{max} = \frac{3\ r_0\omega\tau}{(2 + r_0) + 2\ \{m_0\ (1 + \omega^2\tau^2)\}^{\frac{1}{2}}} \quad (11)$$

Now, deviations of the observed tanΔ_{max} from that calculated by equation 6 (i.e. a tangent deficit or excess) indicate that

the assumptions used in the derivation of the preceding
equations are invalid. For example, Mantulin and Weber (1977)
have shown how strongly anisotropic rotations of aromatic
hydrocarbons result in a maximum tanΔ that is 15-25% smaller
than that expected of isotropic unhindered rotations and
calculated by equation 6. However, even extreme anisotropy
produces tangent defects smaller than 35% (Weber, 1977) and
larger defects can be attributed to either hindered rotations
or to the heterogeneity of fluorophor rotations. The latter
case may be recognised by the broadened distribution of tanΔ
or $\Delta\tau$ against log $2R\tau$ or equivalently temperature, due to
the superimposition of distributions of the individual
components of the mixture.

Recently, Weber (1978) has modified equations 3 and 4 and
6 to allow for isotropic rotations of a fluorophor which have
a lower limit of fluorescence anisotropy (r_∞). In this case
tanΔ can be obtained from :

$$\tan\Delta = \frac{\omega\tau \ (r_0 - r_\infty)(2R\tau)}{1/9 \ m_0 \ (1 + \omega^2\tau^2) + 1/3 S(2R\tau) + m_\infty(2R\tau)^2} \qquad (12)$$

where $m_\infty = (1 + 2 \ r_\infty)(1 - r_\infty)$ and

$$S = 2 + r_0 - r_\infty (4r_0 - 1)$$

The maximal value of tanΔ is now given by :

$$\tan\Delta_{max} = \frac{3\omega\tau \ (r_0 - r_\infty)}{S + 2\{ m_0 \ m_\infty \ (1 + \omega^2\tau^2)\}^{\frac{1}{2}}} \qquad (13)$$

As described , these equations do not permit the estimation
of R and r ∞. However, Lakowicz and Prendergast (1978 a,b)
and Lakowicz *et al.*, (1979) have recently modified these
equations to obtain R by combining data from both differential
phase and steady state polarisation measurements, thus :

$$(m \ \tan\Delta) \ (2R\tau)^2 + (C \ \tan\Delta - A)(2R\tau) + (D \ \tan\Delta - B)$$

$$= 0 \qquad (14)$$

where $A = 3B = \omega\tau(r_0 - r)$; $C = (2r - 4r^2 + 2)/3$

$$D = (m + m_0\omega^2\tau^2)/9 \quad \text{and} \quad m = (1 + 2r) \ (1 - r)$$

In all cases one of the two possible values of R was negative and was therefore ignored. R was substituted in the following equation to obtain r_∞ thus :

$$r_\infty = r + \frac{(r - r_0)}{6R\tau} \qquad (15)$$

Finally the average amplitude ($<\theta>_{max}$) of the angular distribution of the probe around an axis that is normal to the plane of the membrane at times which are long compared with fluorescence lifetime, were obtained (Weber, 1978) from :

$$\cos^2 <\theta>_{max} = \frac{1 + 2(r_\infty /r)}{3} \qquad (16a)$$

$$<\theta>_{max} = \cos^{-1}(\cos^2 <\theta>_{max})^{\frac{1}{2}} \qquad (16b)$$

DYNAMIC STUDIES OF MEMBRANE ADAPTATION

The effect of temperature upon the differential lifetime for liver microsomal and mitochondrial preparations of 5°C and 25°C acclimated green sunfish are presented in figure 14. All preparations exhibited a symmetrical bell-shaped curve which passed through a maximum between 10-15°C, although the curves for mitochondria were somewhat broader and flatter-topped than for microsomes, suggesting a greater heterogeneity in the former. This interpretation is not surprising in view of the fact that mitochondria are composed of two distinct membranes each with a characteristic protein and lipid composition and with a different fluidity (Hackenbrock et al., 1976). Maximum observed values of $\Delta\tau$ were only half of the theoretical values which were calculated assuming isotropic, unhindered rotations by the fluorophor (see table 4). This indicates that assumptions used in steady state studies concerning the rotational characteristics of DPH are certainly not fulfilled in the membranes of fish and rat (data not shown) as well as in artificial membrane systems (Lakowicz et al., 1979). Since there is no reason for believing that the depolarising rotations of DPH are not isotropic then the tangent defect must result either from the combination of different $\Delta\tau$/temperature plots for different populations of DPH or from restrictions to the free rotation of the probe.

   The effect of temperature upon the calculated values of rotational rate (R) and the limiting anisotropy (r_∞) for 5 and 25°C acclimated green sunfish are presented in figures 15 and 16. The Arrhenius plots of R were distinctly curvilinear in all instances and in the mitochondria, at least, the curves

TABLE 4

Maximal differential lifetimes and tangent defects obtained by the phase shift technique for the microsomal and mitochondrial fractions isolated from 5^o and 25^o acclimated green sunfish and rat

Membrane fraction	Source	Temp. (^{o}C)	$\Delta\tau_{max}$ (ns) Obs.	Cal.	Defect (%)	Θ_{max} (15^oC)
Microsomes	5^oC Green sunfish	12^a	0.85^b	1.62^c	47.7^d	44.2^{oe}
	25^o Green sunfish	10	0.67	1.32	49.2	39.4^o
	Rat	18	0.68	1.55	56.1	42.2^o
Mitochondria	5^o Green sunfish	12	0.80	1.50	46.3	44.6^o
	25^o Green sunfish	15	0.65	1.33	50.9	41.0^o
	Rat	18	0.71	1.41	50.0	42.1^o

a Temperature at maximum $\Delta\tau$ obtained from figure 14

b Also obtained from figure 14

c Calculated as described in the 'Methods' section using equation 13

d Calculated as (Observed $\Delta\tau_{max}$/calculated $\Delta\tau_{max}$) x 100

e Calculated as described in equation 16 (methods section using r_∞ at 15° from figure 16)

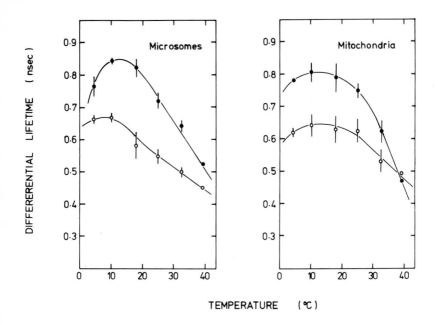

TEMPERATURE   (°C)

Fig. 14. The differential lifetime (Δτ) values of DPH in liver microsomal (a) and mitochondrial (b) fractions of 5°C (●) and 25°C (o) acclimated green sunfish. Values represent the mean ± S.E.M. of 3-7 determinations of a single preparation except for the values at 39°C which was measured only once.

for the differently acclimated fish were identical over the temperature range 4.5 - 39°C. In the microsomes the values of R for the preparation of cold-acclimated fish were lower than those for the warm-acclimated fish, which is contrary to what was expected on the basis of the steady-state experiments.

The limiting anisotropy (r_∞) decreased smoothly with increased temperature in all membrane preparations from both fish and rat (data not shown), indicating a reduced hindrance to probe rotation and, thus, a less ordered membrane interior with increased temperature. In both microsomes and mito-chondria the values of r_∞ for preparations of 5°C acclimated fish were considerably lower than for 25°C acclimated fish. For example, at 18°C, the value of r_∞ for the mitochondria of 5°C acclimated green sunfish was approximately 0.08 whilst the corresponding value for 25°C acclimated green sunfish was 0.128. These results unequivocally demonstrate the greater fluidity of the membranes from cold-acclimated fish compared to warm-acclimated fish. This interpretation

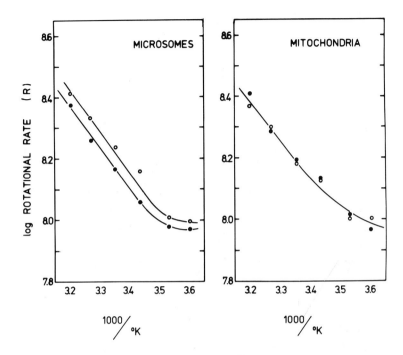

Fig. 15. Rotational rates (R) of DPH in liver microsomal (a) and mitochondrial (b) fractions of green sunfish acclimated to 5°C (●) and to 25°C (o). Values were calculated as described in the text.

relates more specifically to the degree of order of the membrane interior rather than to the rate of probe rotation *per se* and as such provides a clearer insight into the nature of the homeoviscous response than was provided by steady state techniques. Another, perhaps crucial advance offered by this study relates to the magnitude of the structural differences of membranes from 5°C to 25°C acclimated green sunfish. The shift of the r_∞/temperature curve along the temperature axis was approximately 15°C for a 20°C difference in acclimation temperature i.e. a 75% compensation. This is considerably greater than that predicted by steady state techniques and provides a dramatic and, perhaps, more faithful illustration of the efficacy of cellular response to altered cell temperature.

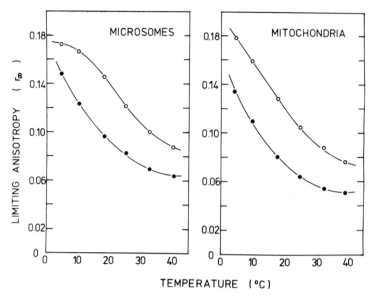

Fig. 16. The limiting anisotropy values (r) of DPH in liver
microsomal (a) and mitochondrial (b) fractions of green
sunfish acclimated to 5°C (●) and to 25°C (o). Values were
calculated as described in the text.

CONCLUSIONS

Advances in the understanding of membrane fluidity and its
adjustments during environmental insult depends critically
upon our detailed knowledge of the dynamic structure of
biological membranes. This in turn depends upon theoretical
and technological advances in the estimation of fluidity by
a variety of spectroscopic techniques. The theory of restricted
rotations in lipid bilayers as elaborated by Israelachvili
et al., (1975) for ESR probes and by Kinosita et al., (1977)
for fluorescent probes, constitutes perhaps the most advanced
descriptive models yet of probe rotation within membranes,
but it is worth pointing out that they are not entirely
satisfactory and more realistic models can be envisioned and
presumably will be elaborated in the future.
    Although the description of membrane fluidity by steady state
polarisation techniques has been superseded by dynamic
fluorescence studies such as those reported here, the former
technique still provides a valid comparative index of membrane
fluidity. Hildenbrand and Nicolau (1979) have recently
compared the two techniques and have concluded that changes
in steady state anisotropy mainly represent changes in the
static properties of DPH; that is, the extent to which the

rotations of the rod-shaped probe are hindered by the anisotropic hydrocarbon chains of the membrane core. A smaller
contribution arises from the rotational relaxation of the
probe within the confines of the rotational cage. However,
the influence of this factor on the steady state anisotropy
also depends upon the fluorescence lifetime, thus confusing
the precise correlation of anisotropy with r_∞ values and
making the steady state parameter somewhat less reliable for
the characterisation of membrane fluidity. The use of microviscosities that are evaluated from steady state measurements
by application of the Perrin equation (Shinitzky and Barenholz,
1978) should thus be considered as semi-empirical measurements,
which, strictly speaking misrepresents the true situation,
but according to Hildenbrand and Nicolau (1979) will not
result in incorrect conclusions regarding the relative
intensity of such molecular motion.
     The steady state technique has the important practical
advantage of relative simplicity, both in instrumentation and
measurement. However, polarisation fluorimeters, whether
commercial or self-built, differ considerably in their
performance and the operating parameters for each individual
instrument should be carefully determined so as to validate
the precision of the fluidity measurements and their compar-
ability with published values. The technology required for
time-resolved measurements is considerably more complex and
less easy to perform. The nanosecond pulse technique provides
a very detailed description of probe rotation but conventional
techniques suffer from the disadvantage of extended acquisition
times, so that extensive analysis of labile biological
preparations, such as mitochondria, are difficult to perform.
The use of tunable laser and synchrotron radiation sources
with very high repetition rates provides the solution to this
problem but these are certainly not easily available. A major
advantage of the differential polarised phase technique is
that measurements may be completed in only 1-2 min and a
simultaneous and detailed temperature scan of a number of
preparations may be carried out within 4-5 hr. In addition,
the calculations required to determine r_∞ and R are relatively
simple and do not require computer facilities.
     To summarise, the studies reported here confirm the existence
of cellular responses in fish that compensate for the effect
of temperature upon the fluidity of their constituent membranes.
The response requires a period of days or weeks to take effect
and thus provide a means of compensating only for seasonal
changes in temperature and not for rapid diurnal temperature
fluctuations. However, homeoviscous responses are not
ubiquitous since some membranous fractions exhibit no or very
restricted fluidity adjustments during thermal acclimation.

The acyl group composition of membrane phospholipids generally becomes more unsaturated during cold acclimation, even in those membranes which show little or no adaptive responses, although in these cases the compositional adjustments are less extensive and are thought to be of a non-functional nature. In a teleological way, the existence of an elaborate homeoviscous response in animal cells testifies to the crucial importance of the degree of fluidity of cellular membranes and to the hypothesis that there is an optimal fluidity towards which the system is constantly being adjusted. Thus, any agent which perturbs membrane fluidity from this optimal condition will elicit homeoviscous adaptation in the addiction of animals to membrane-active anaesthetics and drugs (Nandini-Kishore *et al.*, 1977, 1979). The effect of such drugs is to perturb membrane fluidity and the state of dependence or increased tolerance shown by animals may reflect, at least in part, the compensation of membrane fluidity to its optimal value such that sudden removal of the drug results in a sudden perturbation of membrane fluidity and hence withdrawal symptoms. Another obvious perturbant is the high hydrostatic pressure such as is encountered by a variety of organisms in the ocean depths. Assuming that the membrane fluidity of such organisms is similar to that of surface-living organisms then abyssal organisms must possess an adaptive mechanism that is considerably more impressive than occurs during thermal acclimation.

REFERENCES

Aibara, S., Kato, M., Ishinaga, M.and Kito, M. (1972)
    Biochim. Biophys. Acta 270, 301-306
Andrich, M.P. and Vanderkooi, J.M. (1976), *Biochemistry*
    15, 1257-1261
Cadenhead, D.A., Kellner, B.M.J. and Mueller-Landau, F.
    (1975) *Biochim. Biophys. Acta* 382, 253-259
Caldwell, R.S. and Vernberg, F.J. (1970) *Comp. Biochem.*
    Physiol. 34, 179-191
Cehelnik, E.D., Cundall, R.B., Lockwood, J.R. and Palmer, T.F.
    (1975) *J. Phys. Chem.* 79, 1369-1376
Cossins, A.R. (1976) *Lipids* 11, 306-316
Cossins, A.R. (1977) *Biochim. Biophys. Acta* 470, 395-411
Cossins, A.R., Friedlander, M.J. and Prosser, C.L. (1977)
    J. comp. Physiol. 120, 109-121
Cossins, A.R., Christiansen, J.A. and Prossor, C.L. (1978)
    Biochim. Biophys. Acta 511, 442-454
Cossins, A.R., Kent, J. and Prosser, C.L. (1980) *Biochim.*
    Biophys. Acta 559, 341
Cossins, A.R. and Prosser, C.L. (1978) *Proc. Natl. Acad.*
    Science U.S.A. 75, 2040

Chen, L.A., Dale, R.E., Roth, S. and Brand, L. (1977)
  J. Biol. Chem., 252, 2163-9
Cullen, J., Phillips, M.C. and Shipley, G.G. (1971)
  Biochem. J. 125, 733-642
Dale, R.E., Chen, L.A. and Brand, L. (1977) *J. Biol. Chem.*,
  252, 7500-7510
Esser, A.F. and Souza, K.A. (1974) *Proc. Natl. Acad. Sci.*
  U.S.A., 71, 4111-4115
Fraley, R.T., Jameson, D.J. and Kaplan, S. (1978)
  Biochim. Biophys. Acta 511, 52-69
Friedlander, M.J., Kotchabhakdi, N., and Prosser, C.L. (1976)
  J. Comp. Physiol. 112, 19-45
Fukushima, H., Martin, C.E., Iida, H., Kitajama, Y., Thompson,
  G.A. and Nozawa, Y. (1976) *Biochim. Biophys. Acta* 431, 165-79
Ghosh, D., Lyman, R.L. and Tinoco, J. (1971) *Chem. Phys.*
  Lipids, 7, 173-84
Hackenbrock, C.R., Hochli, M. and Chau, R.M. (1976)
  Biochim. Biophys. Acta 455, 466-484
Hare, F. and Lussan, C.(1977) *Biochim. Biophys. Acta*,
  467, 262-272
Hare, F. and Lussan, C. (1978) *FEBS Letters* 94, 231-235
Hazel, J.R. and Prosser, C.L. (1974) *Physiol. Rev.* 54, 620-677
Hildenbrand, K. and Nicolau, C. (1979) *Biochim. Biophys.*
  Acta 553, 365-377
Hoyland, J., Cossins, A.R. and Hill, M.W. (1979) *J.comp.*
  Physiol. 129, 241-246
Israelachvili, J., Sjosten, J., Erikkson, L.E.G., Ehrstrom,
  M., Graslund, A. and Ehrenberg, A. (1975) *Biochim.*
  Biophys. Acta 382, 125-141
Jacobson, K. and Papahadjopoulis, D. (1975) *Biochemistry*
  14, 152-161
Jameson, D.J., Weber, G., Spencer, R.D. and Mitchell, G.
  (1977) *Rev. Sci. Inst.*, 49, 510-514
Kawato, S.,Kinosita, K. and Ikegami, A. (1977) *Biochemistry*
  16, 2319-2324
Kinosita, A., Kawato, S. and Ikegami, A. (1977) *Biophys. J.*
  20, 289-305
Lakowicz, J.R. and Prendergast, F.G. (1978a) *Science*
  200, 1399
Lakowicz, J.R. and Prendergast, F.G. (1978b) *Biophys. J.*
  24, 213-226
Lakowicz, J.R., Prendergast, F.G. and Hogan,D.(1979)
  Biochemistry, 18, 508-519
Lentz, R.D., Barenholz, Y. and Thompson, T.E. (1976)
  Biochemistry 15, 4521-4528
Mason, R. and Cehelnik, E.D. (1978) *J. Photochem.*, 9, 219-221
Mantulin, W.W. and Weber, G. (1977) *J. Chem. Phys.*,
  66, 4092

Nandini-Kishore, S.G., Kitajima, Y. and Thompson, G.A. (1977) *Biochim. Biophys. Acta* 471, 157-161
Nandini-Kishore, S.G., Mattox, S.M., Martin, C.E., and Thompson, G.A. (1979) *Biochim. Biophys. Acta* 551, 315-327
Miller, N.G.A., Hill, M.W., and Smith M.W. (1976) *Biochim. Biophys. Acta* 455, 644-654
Nozawa, Y., Iida, H., Fukushima, H., Ohki, K. and Ohnishi, S. (1974) *Biochim. Biophys. Acta* 367, 134-147
Penny, D., and Goldspink, G. (1980) *J. Thermal. Biol. (in press)*
Perrin, F. (1926) *J. Phys. Radium* 7, 390-401
Precht, H. (1973) "Temperature and Life", Springer, Berlin
Prosser, C.L. (1964)   "Handbook of Physiology", Sec. 4 (D.B. Dill *et al.*, eds), American Physiological Society, Washington, pp. 11-25
Prosser, C.L. (1973) "Comparative Animal Physiology", Saunders
Salem, L. (1962) *Can. J. Biochem.* 40, 1288-1299
Schreier, S., Polnaszek, C.F. and Smith, I.C.P. (1978) *Biochim. Biophys. Acta* 515, 375-436
Seelig, J. (1977) *Quart. Rev. Biophys.* 10, 353-418
Shinitzky, M. and Barenholz, Y. (1974) *J. Biol. Chem.*, 249, 2652-2657
Shinitzky, M. and Barenholz, Y. (1978) *Biochim. Biophys. Acta* 515, 367-394
Shinitzky, M., Dianoux, A.C., Gitler, C. and Weber, G. (1971) *Biochemistry* 10, 2106-2113
Sidell, B. (1977) *J. Exp. Zool.* 199, 233-250
Sinensky, M. (1974) *Proc. Natl. Acad. Sci. U.S.A.* 71,522-525
Singer, S.J. (1974) *Ann. Rev. Biochem.* 43, 805-834
Singer, S.J. and Nicholson, G.L. (1972) *Science* 175, 720-731
Spencer, R.D. and Weber, G. (1969) *Ann. N.Y. Acad. Sci.* 158, 361-376
Suurkuusk, J. Lentz, B.R., Barenholz, Y., Biltonen, R.L. and Thompson, T.E. (1976) *Biochemistry* 15, 1393-1401
Weber, G. (1977) *J. Chem. Phys.* 66, 4081-4091
Weber, G. (1978) *Acta. Phys. Pol.* A54, 173

DISCUSSION ON DR. COSSIN'S PAPER

Prof. Porter : Have any adaptation experiments of an analogous nature to the ones you describe on fishes been performed on plants ?

Dr. Cossins : I don't know about experiments on plants but certainly on protozoa, *E. coli* and there are differences between pro- and eucaryotes.

Prof. Porter : Is the fluidity of the membrane comparable in plants and animals ?

Dr. Lee : EPR experiments on plants show that the bilayer has the same basic properties as in animals.

Dr. Quinn : Are the effects you describe due to oxygen dissolved in the water instead of temperature since changing the temperature will also change the dissolved oxygen concentration ?

Dr. Cossins : No, I don't believe so. Although experiments of this sort have not been performed on fish, Skriver and Thompson (*Biochim. Biophys. Acta*, (1976) <u>431</u>, 180-188) have eliminated this function in tetrahymena although it may be important in micro-organisms (Brown and Rose, (1969) *J. Bacteriol.*, <u>99</u>, 371-378).

Dr. Smith : What is fluidity ? In NMR experiments fluidity is shown to change down the hydrocarbon chain. What type of average value do you measure with your probes ?

Dr. Cossins : This is an important consideration but unfortunately we don't know precisely what average we measure. DPH does bind to hydrophobic sites and shows no preference for fluid or solid phases (Andrich and Vanderkooi, (1976) *Biochemistry*, <u>15</u>, 1257).

Dr. Johnson : Do the lipid head groups change their packing in the adapted membranes ?

Dr. Cossins : There are no studies on changes in the packing of phospholipid head groups during thermal acclination. In certain cases there are significant changes in the phospholipid head group composition during this type of acclimation (see Cossins, 1976)

Prof. Chapman : Does the fish select the fatty acids which it requires from its diet ? What happens when the diet is deficient in these fatty acids ?

Dr. Cossins : The biosynthetic mechanisms involved in the compositional adjustments in fish are certainly complex and are currently under investigation. In tetrahymena, Thompson and Nozawa (1977) favour a direct viscotropic effect of membrane fluidity on the activity and specificity of desaturases although Lands (*Trans. Biochem. Soc.*,(1980), 8 , 25) believes that there is little evidence of such a 'dialogue' between membranes and their lipid metabolising enzymes. Diet may have an influence on homeoviscous adaptation since it can certainly produce dramatic differences in fatty acid composition and fluidity.

TIME RESOLVED FLUORESCENCE ANISOTROPY STUDIES USING SYNCHROTRON RADIATION

I.H. MUNRO

Daresbury Laboratory,
Warrington WA4 4AD

INTRODUCTION

A complete understanding of natural phenomena inevitably demands a knowledge of kinetic and dynamic processes. Of great interest at the present time are mechanisms for the assembly of atoms in such fundamental processes as chemical reactions, rates of transfer of energy and of the movements associated with specific atomic and molecular moieties in solids, liquids and gases. The time scales concerned are exceptionally short, from about 1 fs (10^{-15} s) to 1 µs (10^{-6} s) or longer. Synchrotron radiation sources with their intrinsic pulsed behaviour have already been applied to the study of time dependent phenomena from the time scale of seconds to picoseconds (10^{-12} s) (Munro and Sabersky,1980). The associated properties of these sources, such as energy, specificity and polarisation afford the opportunity that, given the availability of suitable detectors, every possible electromagnetic radiation induced interaction can be studied as a function of time.

In spectroscopy, measurement of the rate of reorientation of initially oriented states by observing the time dependence of the anisotropy of the emitted fluorescence has led to a measure of the degree and rates of flexibility of large molecules and of the rates of precession of atoms in the presence of an external field. When fluorescence emission is used to identify specific ions, atoms or molecules then the complete time course of any sequence of chemical changes can be in principle identified using the techniques of time dependent emission and excitation spectroscopy. In biology, the mechanisms of photosynthesis, photoreception by the eye, nerve impulse transmission and many others of

fundamental importance in living systems are determined by
molecular-vibrational kinetics on the picosecond time scale.
Of course other dynamic processes such as photochemical
reactions and surface chemical reactions also take place on
this time scale. It is therefore self evident that research
in the picosecond time domain will have a great impact on
science but only when such measurements can be carried out
routinely. To date, timing measurements using synchrotron
radiation have given a resolution of about 50 ps (in the
coincidence experiments) and approximately 5 ps using phase
sensitive detection techniques at high frequencies. These
results have been obtained using storage rings designed for
high energy physics research employing relatively long
(> 100 ps) electron bunch lengths. To seriously contemplate
measurements in the sub-picosecond, i.e. femtosecond time
region, then storage rings must be constructed where relatively large circulating currents (> 1 mA) can be confined
within a bunch length of a few millimetre (i.e. yielding a
pulse duration of <10 ps). The first experimental studies
of bunch length at Stanford, USA, and the specific plan to
maintain short bunch lengths in the BESSY, Berlin, VUV ring
suggest that such criteria will be met in practice within
the near future. The dynamics of molecular systems are at
present limited with lasers largely to studies in the visible
and near UV. However, this work can already be extended
throughout the VUV region using existing synchrotron radiation sources. New experiments which depend on the availability of a high brightness source throughout the VUV
region must have good aberration-free optics available at
all wavelengths before they can be fully exploited. In the
more distant future the use of coherent undulator radiation
combined with the technical advances in multilayer structures for enhanced normal incidence optics may bring a new
range of techniques to bear which to date have been associated with visible and near UV laser physics (European
Science Foundation).
    Since the quality and ultimate time resolution of many
experiments has been restricted by the properties of photon
detectors used, clearly, in the future, new detectors must
be developed which are significantly superior to existing
photomultiplier tubes and to microchannel plate detectors.
The only fast detector immediately in view is the streak
camera which must be driven synchronously with the storage
ring and incorporate a digital data recording system to
accommodate the possibility of carrying out photon counting
studies while retaining the 'bandwidth' of >100 GHz of the
streak tube. A major alternative measuring technique which
is rapidly being advanced is that of phase sensitive

measurements at high frequencies (> 500 MHz). It has already
been shown that, given stability in the operating mode of a
storage ring, an accuracy of a few ps is possible (Sabersky
and Munro,1978). There is no reason why such measurements
should not progress to achieve routinely <100 fs resolution
within the next one or two years.

SYNCHROTRON RADIATION AS A SOURCE FOR TIME-RESOLVED
MEASUREMENTS

The source properties and research applications of synchrotron radiation have been extensively discussed (Munro and
Sabersky, 1980 ; European Science Foundation) and it is
worth summarising them and comparing the properties of
lasers, synchrotron radiation and other incoherent sources
from the point of view of time resolved spectroscopy and
fluorescence anisotropy. This comparison is made in table 1.

The UK Synchrotron Radiation Source (SRS)

At the Science Research Council Daresbury Laboratory, a 2
GeV electron storage ring has been constructed which will
become routinely operational in 1981. This accelerator
(the SRS) will be capable of maintaining up to 1 A of
circulating current with a current 'half-life' of approximately 8 hours. The emission spectrum of the SRS is given
in figure 1. The time structure of the emitted radiation is
determined to a large extent by the radio-frequency used to
accelerate the beam. The radiation is emitted as a series
of well-defined pulses (fwhm $\sim$ 100 ps) with a repetition
rate of 500 MHz. This pulse width corresponds to an electron bunch length of approximately 3.5 cm, and this dimen-
sion can be reduced by alteration of other accelerator
parameters. In the 'single bunch' mode of operation, the
SRS will accelerate only one electron bunch (the SRS is
'filled' when it contains 160 bunches) giving an excitation
pulse of about 100 ps every 320 ns. The SRS should give
< 5×10^7 photons per pulse within a band pass of 1 nm at 200
nm (Lea and Munro, 1980). This pulse intensity of course
will increase as the excitation wavelength is reduced. The
source is ideal for polarisation experiments where the
radiation is almost completely linearly polarised with the
electric vector horizontal (in the plane of the accelerator).

TABLE 1

Comparison of excitation sources for time resolved spectroscopy and fluorescence anisotropy

	Incoherent Sources	Synchrotron Radiation	Lasers
Useful wavelength range	Many different sources are necessary to cover the range >15nm to the infrared	$\sim$0.1 nm to $\sim$ 1 cm	Tunable in ultraviolet and visible. Few lines below 200 nm and many in infrared
Number of photons per pulse (within 0.1% wavelength band)	<10^5 for 1 ns pulse	<10^9 Up to 10^{13} using an undulator for 20ps to 100 ps pulse	$\sim 10^{10}$ in 1 ps pulse of width <10 cm^{-1}
Pulse profile	$\sim$1 ns Wavelength dependent Pulse shape and amplitude often erratic	$\sim$>10 ps Wavelength independent Gaussian and fixed amplitude	$\sim$0.2 ps Wavelength depedent Gaussian and slightly variable amplitude
Pulse repetition rate	DC to <100 MHz	$\sim$1 MHz to 500 MHz	DC to $\sim$100 MHz
Source size and divergence	Few mm^2 isotropic, incoherent and unpolarised	$\sim$1 mm^2 <10 milliradians incoherent, 100 % linearly polarised in machine plane	$\sim$1 mm^2 <1 milliradians, coherent and polarised

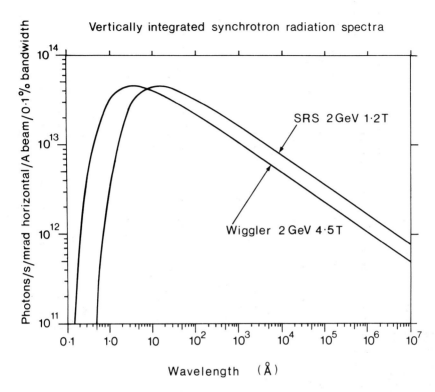

Fig. 1. Spectral curves from a normal bending magnet (1.2 T) and from a 4.5 T wiggler magnet for a 2 GeV 1A beam in the SRS (Lea and Munro, 1980)

TIME RESOLVED FLUORESCENCE ANISOTROPY AND SPECTROSCOPY OF LARGE MOLECULES

An excellent illustration of the merits of synchrotron radiation as an excitation source for time-resolved fluorescence polarisation spectroscopy is provided in its recent application to the study of proteins (Munro, Pecht and Stryer, 1979). The existence of a fluorescent chromophore located at a known site within any molecule can be used to derive a considerable amount of information relating to the micro-environment of that specific site. Energy-transfer and quenching studies measured via changes in the fluorescent lifetime of the chromophore can provide a 'spectroscopic ruler' (Yguerabide, 1972) for proximity relationships within the molecule and yield information about the diffusion of molecules such as oxygen into the interior of the protein.

When linearly polarised light is used for excitation and the
degree of polarisation of the emission is studied we have the
possibility of monitoring geometrical relationships on the
same short time-scale. This technique has so far been
largely confined to the use of extrinsic probes which have
been chemically attached to the molecule under investigation.
Important information has been derived using such extrinsic
probes, for example, in the estimation of the shape, size and
mobility of very large molecules. However, such data may be
criticised because of possible modifications introduced by
the attachment of the fluorescent probe either to the structure or the biological activity of the host molecule. The
ideal chromophore in the study of proteins is given by one
of its constituent amino-acids used as an intrinsic probe.
Unfortunately only three different amino acids are fluorescent which possess modest quantum yields, short lifetimes
and absorb and emit at rather short (< 300 nm) wavelengths.
In order to perform a useful analysis of any data from time-
resolved anisotropy measurements using a fluorescent amino-
acid, the host molecule must - ideally - be known to contain
only a single chromophore molecule. The work described here
was concerned primarily with a number of different proteins
containing a single tryptophan residue. Of course, when
several trytophans are associated with the emitted fluorescence, analysis of the motion of individual residues is
exceedingly difficult because the different time-dependent
anisotropy profiles are superimposed. Even with only a single
tryptophan residue present, it is necessary to set the
excitation wavelength at the extreme long wavelength edge of
the absorption spectrum. Because the position of the edge
and the spectral profile are often dependent on the solvent
environment of the residue, an intense tunable polarised
source is essential for the success of this work.
    Pulsed synchrotron radiation from the Stanford storage ring
SPEAR was used as the excitation source in these studies.
The horizontal divergence of the beam from SPEAR was about
1.8 milliradians and the linearly polarised radiation was
delivered to the sample with the electric vector vertical
following two reflections. The spectral bandpass used for
tryptophan excitation was usually about 2 nm and the emitted
fluorescence was collected at 90° to the excitation beam
through a rotatable polarising filter. The emission decay
kinetics were studied using a single photon counting coincidence apparatus shown in figure 2. The excitation pulse
width was approximately 600 ps (limited solely by the photomultiplier) and the pulse repetition rate was 1.28 MHz. The
importance of synchrotron radiation in this context is that

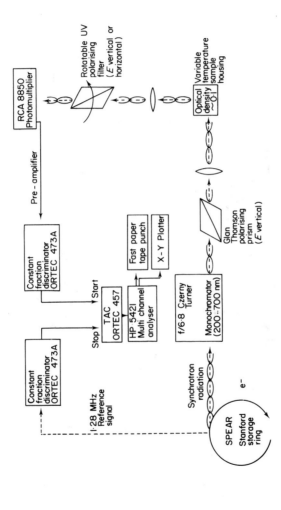

Fig. 2. Schematic arrangement of apparatus used for time resolved fluorescence anisotropy experiments using radiation from the Stanford storage ring, SPEAR.

the extreme stability - in shape and amplitude - of the
excitation pulse is combined with a very high repetition
rate. In the case of most of the proteins studied, over
10^5 peak channel counts would be accumulated taking from ten
to forty minutes. Such relatively short measuring times are
important where sample integrity is difficult to establish
and maintain as is the case with many proteins and particularly so at high temperatures. A further consideration is
that the time position and profile of the pulse is independent of wavelength for synchrotron radiation and therefore,
unlike any other source, the instrument response function
can be properly measured at the sample emission wavelength.
The introduction of an adjustable time-shift in fitting
experimental data to compensate for the time/wavelength
dependence of normal sources is particularly hazardous in
anisotropy measurements since the initial anisotropy (A_0)
can be either positive or negative. The high pulse intensity
is important since measurements can then be made at rather
low concentrations (below 10^{-6} M) approaching the conditions
which might be found in real systems.

ANISOTROPY THEORY AND ANALYSIS

The rate at which fluorescence emission polarisation becomes
random as a result of rotational Brownian motion depends on
the degree of flexibility of the fluorescent chromophore with
respect to the macromolecule and on the external size, shape
and also any internal movement of the macromolecule (Wahl,
1975). Such motions can be studied by measuring the intensities of fluorescence polarised horizontally, h(t) and
vertically v(t) as a function of time.
  The total fluorescence intensity F(t) and the fluorescence
emission anisotropy are given by

$$F(t) \;=\; v(t) + 2h(t) \;=\; F_0 e^{-t/\tau}$$

$$A(t) \;=\; \{v(t) - h(t)\}\,/F(t) \;=\; A_0 e^{-t/\phi}$$

F_0 is the initial fluorescence intensity, τ the fluorescence
lifetime, A_0 the initial anisotropy ($-0.2 < A_0 < 0.4$) and ϕ
the rotational correlation time. In practice. the experimentally measured parameters $v_{exp}(t)$ and $h_{exp}(t)$ are
distorted by the response function of the apparatus, I(t),
where

$$v_{exp}(t) \;=\; \int_0^\infty I(t')v(t-t')dt'$$

The influence of the apparatus must be taken into account in order to calculate excited state lifetimes, rotational correlation times or amplitudes. For many materials the emission anisotropy and excited state kinetics cannot be represented by single exponential functions. This occurs, for example, if the chromophore has more than one mode of flexibility, when the macromolecule is non-spherical and when the chromophore experiences more than one local environment during its excited state lifetime. In these circumstances both the fluorescence intensity F(t) and the emission aniso-tropy A(t) are given by the sum of exponential terms with differing amplitudes and τ and ϕ values. Present computing procedures can usefully identify up to three components in F(t) and A(t) although this will apply only to statistically good data with well separated component slopes.

The results of making measurements with a 'model' compound are illustrated in figure 3. A series of glycerol-water mixtures were used as solvent for N-acetyl tryptophamide and the rotational correlation times (ϕ) derived for the tryptophan chromophore. As the solvent viscosity was reduced from 1445 cP to 1.4 cP, ϕ decreased from 16.8 ns to about 50 ps as predicted on the basis of a spherical molecular rotor. The raw anisotropy data are shown alongside aniso-tropy plots which were calculated by convolution of the instrument response function with an excited state lifetime of 5.0 ns, an initial anisotropy (A_0) of 0.2 and a range of correlation times (ϕ) from 100 ps to 10 ns. The synchrotron radiation source properties are particularly important when $\phi < 0.5$ ns and the best fit is established by the time posit-ion rather than the slope of the anisotropy decay plot. Good statistics (a maximum number of counts), a negligible amount of scattered light (which would severely distort A_0 and small values of ϕ) and a minimum excitation pulse duration are essential before reliable anisotropy data can be obtained, particularly in subnanosecond time region.

Figures 4 and 5 reveal the range of information which can be elucidated from anisotropy measurements when they are interpreted along with other spectral, structural and chem-ical information. In figure 4, the tryptophan residue is in basic myelin protein and in nuclease B - proteins with rather similar molecular weight - exhibit widely different rotational correlation times. In nuclease B the tryptophan undergoes a single mode of rotational motion with a correlation time (ϕ = 9.8 ns) closely similar to the value of 7.6 ns calculated for a rigid hydrated sphere at the same temperature. This result is consistent with x-ray structural information which indicates that in nuclease A, the residue is hydrogen-bonded

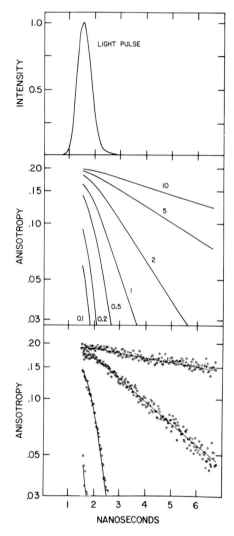

Fig. 3. Fluorescence anisotropy data (from Munro, Pecht and Stryer, 1979). The full instrument response function measured at 340 nm is shown with fwhm of 650 ps. This function was used to derive the simulated fluorescence emission anisotropy decay curves assuming a range of rotational correlation times from 0.1 ns to 10 ns. The raw anisotropy data for 10 μM N-acetyltryptophanamide at 20°C yield correlation times of 16.8 ns (top curve), 3.7 ns, 0.38 ns and ∿0.05 ns for glycerol water mixtures of 1445, 304, 17 and 1.4 cp respectively.

and therefore fixed to the entire protein. In myelin basic
protein, the very short correlation times (ϕ= 90 ps and 1.3
ns) can be interpreted as showing exposure of the tryptophan
residue to the solvent. These results are supported by other
data which indicate that basic myelin protein behaves as a
random coil.

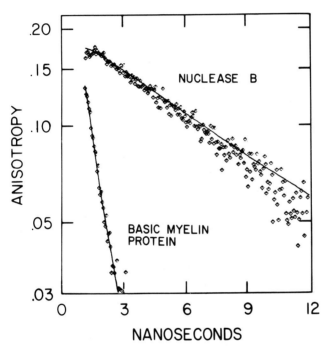

*Fig. 4.   The fluorescence emission anisotropy decay curves
taken from Lea and Munro, 1980 for s.aureus nuclease B and
basis myelin protein. The rotational correlation times are
9.85 ns for nuclease B and 0.09 ns, 1.26 ns for basis
myelin protein.*

In figure 5, two quite distinct kinds of rotational motion
are evident in the emission anisotropy kinetics of holo- and
apoazurin. In the natural protein, the longer correlation
time (ϕ= 11.8 ns) corresponds well to the rotational period
of the complete molecule. This time is reduced (ϕ= 5.8 ns)
when the active-site copper ion is removed showing that apoazurin has modes of flexibility not exhibited by holoazurin.
In holo- and apoazurin a fast rotation is observed, which

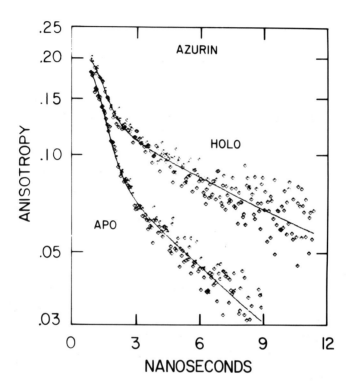

*Fig. 5. The fluorescence emission anisotropy decay curves
of the holo- and apo- protein forms of Pseudomonas
aeruginosa azurin taken from Lea and Munro, 1980. The best
fit parameters for holoazurin give correlation times of 0.51
ns and 11.8 ns and those for apoazurin give 0.49 ns and 6.84
ns respectively. The fluorescence lifetimes were measured
to be 0.75 ns and 4.15 ns for holoazurin; 0.88 ns and 4.79
ns for apoazurin.*

when associated with fluorescence spectral data can be interpreted as motion of the tryptophan residue within the interior
of the protein. If the fluorescence transition moment of the
tryptophan attached to the polypeptide chain is assumed to
rotate freely within a cone and also that the entire azurin
molecule is rotating within the solvent, then the amplitudes
associated with each of these motions can be used to derive
a measure of the angular range of the motion of the residue
within the azurin (Munro, Pecht and Stryer, 1979). The

results yield a large angular range for this motion (more than 40^0) which is greater for apo- than for holoazurin.

SUMMARY

So far, relatively little information has been presented concerning the variations of amino acid mobility within different solvents and molecular sites. Furthermore, there are only two intrinsic probes (tyrosine and tryptophan) which can be readily used and whose behaviour can be interpreted. Even tryptophan data must be treated with caution to avoid the possibility of purely molecular photophysical effects being interpreted as rapid initial changes in anistropy. In this respect, tyrosine should prove a more photophysically reliable chromophore. Despite such reservations, it is obvious that a study of the fluorescent amino acids will lead to a better understanding of protein flexibility in regard to a wide range of biological functions. Indeed, it is likely that rapid structural fluctuations are the essential fundamental steps in structural changes associated with such functions. Comparison between a number of similar proteins containing the same single residue should lead to a correlation of amino acid sequence and chromophore site with chemical or biological function. The technique of time-resolved fluorescence anisotropy has also been used to observe temperature effects and denaturation on the specific chromophore site. Although not discussed here, all measurements of anisotropy include measurements of fluorescence lifetimes and of relative quantum efficiency. Quite independently of time-dependent changes in geometry, the data will yield an indication of environment changes via fluorescence quenching and perhaps give an indication of residue location using energy transfer calculations.

In the general case, macromolecules will contain more than one fluorescent residue. In these circumstances it may not be possible to attribute very specific geometrical properties to a residue on the basis of time-dependent anisotropy studies unless chemical or structural knowledge or techniques are available which allow us to reject or to block information from all but one emitting site. Even when several residues are responsible for the fluorescence emission, for example from the several tryptophan residues present in immunoglobulins and in immunoglobulin fragments, it is clear that the structural flexibility of these large molecules can be roughly characterised and that whole molecule rotation, domain mobility and residue rotation can be separately identified.

When new extrinsic probes are developed, they will no doubt
be used to observe conformation changes in DNA in different
solvent environments, to study RNA polymerases, membrane
structure and properties and other systems; all in time
scales extending down to the few picosecond region which is
the basic time scale for many molecular assembly or rearrangement mechanisms and therefore the ultimate goal for
routine studies using these techniques.

REFERENCES

European Science Foundation (1 Quai Lezay-Marnesia,
    67000 Strasbourg, France). 'The case for a European VUV
    storage ring'.
Lea, K.R. and Munro, I.H. (1980), 'The Synchrotron Radiation
    Source at Daresbury Laboratory'.
Munro, I.H., Pecht, I. and Stryer, L., (1979), *Proc. Natl.
    Acad. Sci. U.S.A.* 76, 55-60.
Munro, I.H. and Sabersky, A.P. (1980) in 'Synchrotron
    Radiation Research', (H. Winick and S. Doniach, eds),
    Plenum Press, New York.
Sabersky, A.P. and Munro, I.H. (1978) in 'Picosecond
    Phenomena', (C.V. Shank, E.P. Ippen and S.L. Shapiro, eds)
    Springer-Verlag, Berlin, pp. 85-88.
Wahl,P. (1975), *New Tech. Biophys. Cell. Biol.*, 2, 233-241.
Yguerabide, J. (1972), *Methods Enzymol.*, 26, 498-578.

DISCUSSION ON DR. MUNRO'S PAPER

Dr. Beddard : Can one separate the correlation times you observe i.e. is there a coupling between rotational motion of the tryptophan and motion of the protein ?

Dr. Munro : We used the analysis of Kinosita *et al.*, (*Biophys. J.*,(1977), 20, 289) assuming uniform motion in the cone and we can separate the two motions i.e. that of the local motion of the probe from that of the whole protein.

Dr. Dale : How is tumbling of the whole body related to tumbling of a part ? Coupling depends upon the cone angle of tumbling.

Dr. Munro : The rotation times are clearly very different (fig. 5) and so one can easily separate the two types of rotation.

Dr. Bayley : The fluorescence in azulin is evidently structured. How can one reconcile this with a large angular range for this rotation and how can one relate this to NMR measurements ?

Dr. Dale : One possibility that might be worth considering is that two sites exist and the probe (tryptophan) shuttles between these but in both sites the spectrum is structured.

Dr. Munro : Yes, the multiple site model is attractive. Possibly one could obtain extra information from detailed fluorescence decay studies.

Dr. Devonshire : Can one distinquish the tryptophan emission depolarisation arising from energy transfer from that resulting from rotational motion in multi-tryptophan systems?

Prof. Porter : As Dr. Teale has demonstrated (this volume) changing the temperature can affect the energy transfer and rotational motion in different ways.

Dr. Teale : Do you intend to look at tyrosine-containing proteins without tryptophan ? They have a number of advantages such as invariant emission wavelength and there is no change in the direction of the transition dipole in going to the excited state.

Dr. Munro :   We are presently looking at tyrosine-containing compounds.

Dr. Dale : Referring back to possible energy transfer in your experiments, was the excitation wavelength at the red edge of the absorption spectrum ?

Dr. Munro : Yes, at the red edge.

Dr. Dale : Then it is unlikely that energy transfer occurs since it has been shown that the resulting depolarisation which occurs on excitation over the main part of the absorption spectrum in the presence of of energy transfer does not occur for long wavelength excitation (the Weber red edge effect - Weber and Shinitzky, *Proc. Natl. Acad. Sci. U.S.A.,*(1970), <u>65</u>, 823)

Prof. Porter : What are the advantages of the synchrotron radiation source over a laser ?

Dr. Munro : The main advantage is the extremely wide tunability of the SRS. In addition, the identical wavelength profile at all wavelengths is an advantage over other sources when convolution is required for data analysis. With slight modifications to the experimental procedure, one can contemplate experiments on time-resolved X-ray crystallography or for instance, far infrared rotation experiments.

DIPHENYL HEXATRIENE AND SOME DERIVATIVES
AS FLUORESCENT PROBES OF MEMBRANE STRUCTURES

R.H. BISBY, R.B. CUNDALL, L. DAVENPORT,

I.D. JOHNSON and E.W. THOMAS

Department of Biochemistry
University of Salford
Salford M5 4WT

1,6-Diphenylhexatriene (DPH) has several favourable qualities
(Shinitzky, 1978) as a probe of membrane fluidity : its
fluorescence emission has a high quantum yield, and is well
separated from the absorption bands of membrane constituents.
In addition, DPH is stable chemically and is easily incorporated into most membranes. The photophysical behaviour of
DPH in both organic solvents and liposomes has been studied
in detail (Kawato *et al.*, 1977; Chen *et al.*, 1977) although
the lack of clear interpretation of the data is rather
unfortunate. Finally, a wide range of membrane types have
now been studied using DPH, facilitating their rapid comparison. This work describes some typical applications of DPH
as a probe of bilayer structure.

In the first of these applications the association of
urushiol (the poison ivy allergen) and some analogues, with
bilayer membranes was investigated. Urushiol is a complex
mixture of n-alk(en)yl catechols, the alkyl chain length
varying from C_{15} to C_{17}. A high proportion (95%) of chains
are unsaturated, being either mono, di-, or tri-olefinic
(Gross *et al.*, 1975). In this study, pentadecylcatechol
(PDC) has been used as a model for urushiol. Both PDC and
the components of urushiol are lipophilic compounds and
would be expected to partition into biological membranes.
Whether or not the allergenic properties of urushiol are
associated with its possible interaction with membranes
remains unknown, but interest in the interaction arises from
the observation that alkylcatechols can modify the immuno-
genicity of tumour cells (Baldwin *et al.*, 1979; Byers and
Baldwin, 1979). Figure 1 shows the effect on DPH fluorescence

of adding PDC and urushiol to single bilayer vesicles of dipalmitoylphosphatidylcholine (DPPC). DPH is known to report the liquid crystalline to fluid phase transition in DPPC at $\sim$ 41°C by a sharp drop in fluorescence polarisation (Andrich and Vanderkooi, 1976).

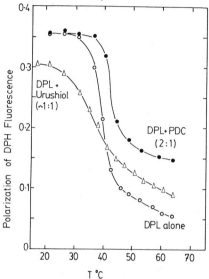

Fig. 1. Polarisation of DPH fluorescence as a function of temperature in membrane vesicles of DPPC containing either PDC or urishiol. Total lipid concentration $\simeq$ 1 mM in phosphate buffered saline, pH 7.4

Addition of PDC in a high molar ratio to the PPC (PDC:DPPC = 1:2) results in an increased polarisation of DPH fluorescence at temperatures above the phase transition. In this respect, PDC behaves very similarly to cholesterol in rigidifying the fluid membrane. At temperatures below the phase transition, PDC has no effect on membrane fluidity as reported by DPH, presumably because of the similarity in chain length of the alkyl groups of PDC and DPPC. Urushiol behaves somewhat differently from PDC, in increasing the fluidity of the liquid crystalline phase below 40°C and tending to obscure the phase transition. This difference in behaviour between urushiol and PDC can be ascribed to the unsaturation in the alkenyl chains of the former compound. Such unsaturation is well known to prevent formation of the crystalline array of alkyl chains in lipids below the phase transition (Seelig, 1978). Experiments with egg lecithin vesicles (figure 2a), which exist in the fluid state over the temperature range studied, also show that PDC has a

rigidifying effect on the fluid membrane state.
   Having found that the interaction of both urushiol and an
analogue with artificial lipid bilayers could be demonstrated
using DPH as a probe, the interaction was also investigated
with the erythrocyte membranes (figure 2b), the polarisation
of DPH was found to decrease implying an increase in fluidity
caused by the uptake of PDC. This is opposite to the effect
of PDC in the two artificial membranes described above.

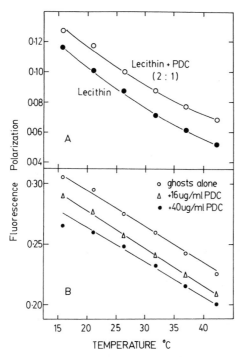

*Fig. 2. Effects of adding PDC to membranes, as measured by
polarisation of DPH fluorescence. (A) Lecithin vesicles,
(B) erythrocyte ghost membranes*

The erythrocyte ghost membrane contains proteins and cholesterol, in addition to phospholipid (cholesterol: PL ∿
0.8 : 1). Additional experiments show that the cholesterol
content of the erythrocyte membrane can account for this
result. Figure 3 shows that in mixed lecithin-cholesterol
vesicles as the cholesterol:lecithin ratio is raised, the
modulating effect of another urushiol analogue, dodecylresorcinol, is diminished and eventually, at an equimolar
ratio is reversed. This demonstrates the important contribution cholesterol makes to the dynamics and structure of
biological membranes. Rather similar effects have been

Bisby *et al.*

observed by Pang and Miller (1978) using a spin probe to
investigate the effects of lipophilic drugs on lecithin-
cholesterol membranes.

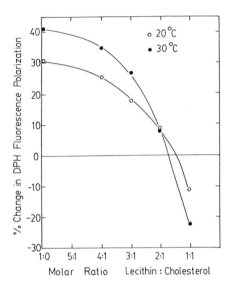

*Fig. 3. Percentage changes in the polarisation of DPH
fluorescence in mixed cholesterol/lecithin vesicles on
incorporation of dodecylresorcinol*

DPH has also been used as a fluorescent probe in the examination of the effects of ionizing radiation upon the erythrocyte ghost membrane (Purohit *et al.*, 1980). Figure 4 shows
that if DPH is added to the membrane suspension following
irradiation, damage to the membranes is revealed as an
approximately exponential decrease in the fluorescence
intensity with increased radiation dose. That the response
is dependent on the presence of free radical scavengers
(such as N_2O, CO_2 and O_2) shows the damage is due to an
indirect mechanism, involving free radicals produced by
radiolysis of the aqueous phase. The increase in radio-
sensitivity in N_2O saturated suspensions compared with the
N_2-saturated suspensions show that the ·OH radical is mainly
responsible for the observed effect ; N_2O converts the
hydrated electron (e_{aq}^-) to ·OH radical, thereby doubling the
yield to the latter species :

$$e_{aq}^- \ + \ N_2O \ \xrightarrow{H_2O} \ \cdot OH \ + \ OH^- \ + \ N_2$$

The further increase in damage observed in oxygenated

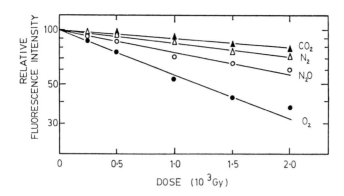

Fig. 4. Fluorescence intensity of DPH added to γ-irradiated erythrocyte ghost membrane suspensions (0.15 mg/ml protein) at pH 7.4

suspension must be due to reaction of O_2 with a lipid radical, R.,

$$R\cdot + O_2 \xrightarrow{R\cdot} RO\cdot_2$$

since the measured yield of lipid peroxides is also highest in the oxygenated suspension. Further studies with other scavengers of the radical species involved also demonstrates a relationship between the structural damage, measured using fluorescent probes, and peroxidative damage. In the intact cell, this damage is finally expressed as an increased permeability of the membrane to ionic species, and finally haemolysis occurs (Myers, 1970).

A similar decrease in the fluorescence intensity of 1-anilinonaphthalene-8-sulphonate (ANS) is observed on irradiation of the erythrocyte membranes. However, the response of the polarisations of DPH and ANS fluorescence to increasing dose are different. Figure 5 shows that a small increase in DPH fluorescence polarisation occurs, which, although a full time-resolved study has not yet been undertaken, would imply that radiation damage decreases the fluidity of the membrane. This would be consistent with a peroxidative mechanism which selectively acts upon the unsaturated fatty acyl chains of the membrane phospholipids. In contrast, the polarisation of ANS fluorescence decreases markedly. It is thought that the different responses of these two fluorescence probes is due to their occupying different regions of the membrane structure. Whilst ANS is thought to probe the membrane-aqueous interface, DPH probes the hydrophobic interior of the membrane. Hence this study emphasises the need to have available fluorescent probes which can be positioned at varying locations in the membrane structure.

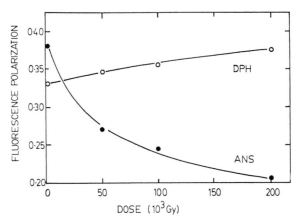

Fig. 5. Fluorescence polarisation of DPH and ANS added to erythrocyte ghosts (0.3 mg protein/ml) after irradiation in O_2-saturated suspensions. Measured at 10°C.

Despite the advantages noted above, DPH has some drawbacks as a membrane probe. Firstly, its high lipophilicity ensures that DPH partitions into both rigid and fluid lipid regions (Lentz, 1976) of membranes, so that only an averaged value of bilayer properties is reported. An assessment of lipid microheterogeneity in natural membranes would be extremely valuable. Again, the emission maximum of DPH is insensitive to solvent polarity and viscosity (Cehelnik *et al.*, 1975), meaning that extra information available (in principle) from the probe is not being reported. Finally, DPH may undergo appreciable photochemical changes : trans → cis isomerisation, dimerisation, and reaction with solvent molecules can be envisaged. We have evaluated a range of DPH derivatives (Cundall *et al.*, 1979) as possible membrane probes, in the hope of obtaining probes with improved characteristics.

All derivatives were prepared by the Wittig reactions outlined in figure 6. The high chemical stability of the diphenylhexatrienyl system also enables several chemical operations (e.g. reduction of nitro functions to amino) to be carried out. Structures, absorption and emission char- acteristics of selected derivatives are shown in tables 1 and 2. Several derivatives (table 3) show striking solvent emission shifts. A series of DPH derivatives is thus readily available, where the emission properties differ quite markedly to those of DPH itself. Although many compounds show

TABLE 1

Structures of DPH derivatives

DNPH Di-nitro analogue MW 322	M2°PH Mono-dimethylammonium analogue MW 275
NO_2—⟨⟩—=-=-=—⟨⟩—NO_2	⟨⟩—=-=-=—⟨⟩—$N(Me_2)$
MNPH Mono-nitro analogue MW 277	M3°PH Mono-trimethylammonium analogue MW 417
NO_2—⟨⟩—=-=-=—⟨⟩	⟨⟩—=-=-=—⟨⟩—$\overset{+}{N}(Me_3)$ I^-
MNPYH Mono-nitro, monopyridyl analogue MW 278	DPYH (N3 isomer) Dipyridyl analogue MW 234
NO_2—⟨⟩—=-=-=—⟨N⟩	⟨N⟩—=-=-=—⟨N⟩
D 3°PH Di-trimethylammonium analogue MW 419	DAPH Di-acetamido analogue MW 346
$(Me_3)\overset{+}{N}$—⟨⟩—=-=-=—⟨⟩—$\overset{+}{N}(Me_3)$ Cl^- Cl^-	A_cHN—⟨⟩—=-=-=—⟨⟩—NHA_c
MNPYH Mono-nitro, monopyridyl analogue MW 278	DPYH (N3 isomer) Dipyridyl analogue MW 234
NO_2—⟨⟩—=-=-=—⟨N⟩	⟨N⟩—=-=-=—⟨N⟩
D 3°PH Di-trimethylammonium analogue MW 419	DAPH Di-acetamido analogue MW 346
$(Me_3)\overset{+}{N}$—⟨⟩—=-=-=—⟨⟩—$\overset{+}{N}(Me_3)$ Cl^- Cl^-	A_cHN—⟨⟩—=-=-=—⟨⟩—NHA_c
DCPH Di-cyano analogue MW 282	MAPH Mono-acetamido analogue MW 289
CN—⟨⟩—=-=-=—⟨⟩—CN	⟨⟩—=-=-=—⟨⟩—$NHAc$

TABLE 2

Table of spectral maxima of DPH derivatives

	Solvent	λ_{max} (nm)	E_{max} ($M^{-1} cm^{-1}$)	λ_{em} (nm)	pK
MNPH	Toluene	405	46,000	510	-
MNPH	Aceto-nitrile	405	46,000	650	-
DNPH	Toluene	410	59,000	495	-
DNPH	Aceto-nitrile	410	59,000	610	-
MAPH	Chloroform	368	65,000	445	-
DAPH	Methoxy-ethanol	380	87,000	450	-
DCPH	Hexane	372	86,000	455	-
M3PH	Methanol	354	53,000	440	-
D3PH	Methanol	356	49,000	435	-
M2PH	Acetone	390	52,000	500	3.7
MNPYH-2	Acetone	384	51,000	525	3.1
MNPYH-3	Acetone	384	49,000	525	3.7
DPYH-3	Methanol	350	57,000	430	3.8
DPYH-4	Methanol	352	-	430	6.1

(λ_{max} - absorption maximum ; E_{max} - extinction coefficient ; λ_{em} - fluorescence emission maximum)

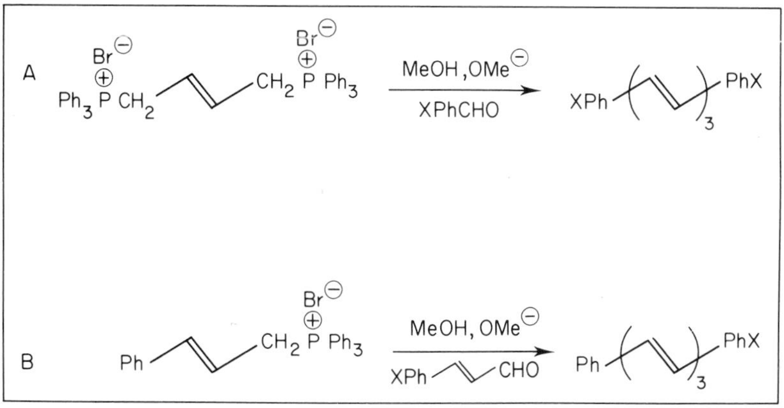

Fig. 6. Synthetic routes to symmetrical (A) and unsymmetrical (B) analogues of DPH

reasonable quantum yields in non-polar solvents, the general trend in all derivatives is towards low quantum yields and short fluorescence lifetimes. Several lifetime measurements (obtained using a synchrotron radiation source - SPEAR at Stanford) are summarised in table 4 .

TABLE 3

Emission maxima of some DPH derivatives (nm)

Compound	$\lambda_{emission}$ (nm)	
	Hexane	Methanol
MNPH	435, 455 (λ_{ex} 400)	650 (λ_{ex} 405)
M2PH	428, 454 (λ_{ex} 390)	530 (λ_{ex} 390)
MC2PH*	468, 500 (λ_{ex} 420)	646 (λ_{ex} 420)

* 4-dimethylamino,4'-cyano-DPH

An attempt has been made to assess the locations of the more polar probes in lipid bilayers using 5- and 16-doxyl stearic acids as quenchers. The possibility exists that the more polar probes (e.g. 4,4'-dicyano DPH and 4,4'-diacet-amido DPH) could adopt significantly different orientations in the bilayer as compared to DPH. A probe molecule orien-tated parallel to the bilayer plane near the bilayer surface would be expected to be quenched more efficiently by 5-doxyl

TABLE 4

Quantum yields, ϕ_f, and lifetime, τ, of the analogues of DPH in various solvents

	Acetonitrile	Benzene/Toluene	Hexane	Hexanol/Butanol	Methanol/Ethanol	Chloroform
DPH			$\phi = 0.8$ $\tau = 12.1$		$\tau = 5.8$	$\tau = 6.5$
DAPH	$\phi = 0.18$ $\tau = 1.1$	$\tau = 0.84$	$\tau = 0.37$	$\phi = 0.61$ $\tau = 2.0$	$\tau = 1.5$	$\phi = 0.57$ $\tau = 1.7$
DNPH	$\phi = 0.3$ $\tau = 2.2$	$\phi = 0.1$ $\tau = 0.3$			$\tau = 0.49$	$\tau = 1.9$
DYPH					$\phi = 0.07$ $\tau = 1.3$	
D3°PH					$\phi = 0.04$ $\tau = 0.51$	
MNPH	$\phi = 0.1$ $\tau = 1.5$	$\phi = 0.2$ $\tau = 0.82$	$\phi = 0.1$ $\tau = 2.4$			$\tau = 1.6$
MNPHY	$\phi = 0.18$	very weak emission			very weak emission	$\tau = 2.2$

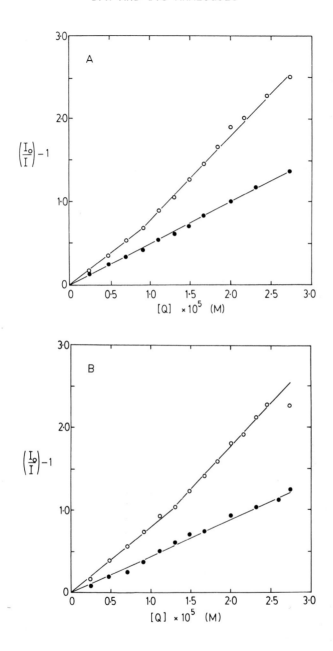

Fig. 7.   Quenching of DCPH (O) and DPH (●) fluorescence
in egg lecithin liposomes at 25°C by (A) 5-doxyl stearic
acid and (B) 16-doxyl stearic acid

stearic acid than by 16-doxyl stearic acid. Stern-Volmer
plots for quenching of DPH and its 4,4'-dicyano derivative
are shown in figures 7a and 7b. The probes appear to be
quenched in a similar way by both 5- and 16-doxyl stearates
indicating that their orientations in the bilayer are
probably identical (the lifetime difference between the probes
probably accounts for the difference in the slopes of the
quenching curves). Studies on the partition of the various
probes into liposomes of defined structures are in progress.
    The photophysical basis of the large solvent emission
shifts seen for some probes is also under investigation
(table III). These probes, showing short-lived emissions,
are convenient for monitoring lipid fluidity by intensity
measurement alone. Thus, 4-dimethylamino-DPH shows a
dramatic decrease in emission intensity over the phase
transition in dimyristoyl lecithin bilayers. Such effects
can be used to study probe transfer between different
vesicle types and to study effects of perturbants on bilayer
structure. The extreme sensitivity of the emission characteristics of the probes shown in table III to solvent
polarity and rigidity indicate their suitability as probes
of membrane lipid heterogeneity.

ACKNOWLEDGEMENTS

Measurements of fluorescence polarisation were performed with
an instrument constructed with a grant from the Royal Society.
We wish to thank Professor R.W. Baldwin for helpful
discussions and the gift of urushiol. Some of this work
was supported through S.R.C. studentships to I.J. and L.D.
    Some of the materials incorporated in this work were
developed at the Stanford Synchrotron Radiation Laboratory
which is supported by the National Science Foundation (under
contract DMR 78-27-489), in cooperation with SLAC and the
Department of Energy.

REFERENCES

Andrich, M.P. and Vanderkooi, J.M., (1976) *Biochemistry*,
    15, 1257-1261
Baldwin, R.W., Embleton, M.J. and Pimm, M.V. (1979) *In*
    "Antiviral Mechanism in the Control of Neoplasia"
    (P. Chandra, ed.). pp. 333-353. Plenum, New York
Byers, V.S. and Baldwin, R.W., (1979) *Symp. Mo. Cell. Biol.*
    16 *(TB Lymphocyte Recognition Function)* 603-622
Cehelnik, E.D., Cundall, R.B., Lockwood, J.R. and Palmer,
    T.G., (1975), *J. Phys. Chem.*, **79**, 1369-1376

Chen, L.A., Dale, R.E., Roth, S., and Brand, L. (1977)
  J. Biol. Chem, 252, 2163-2169
Cundall, R.B., Johnson, I.D., Jones, M.W., Thomas, E.W., and
  Munro, I.H. (1979) *Chem. Phys. Letters*, 64, 39-42
Gross, M., Baer, H. and Fales, H.M. (1975) *Phytochemistry*
  14, 2263-2266
Kawato, S., Kinosita, K. and Ikegami, A. (1977) *Biochemistry*,
  16, 2319-2324
Lentz, R.B., Barenholz, Y. and Thompson, T.E. (1976)
  Biochemistry , 15, 4521-4528
Myers, D.K. (1970) *Adv. Biol. Med. Phys.*, 13, 219-230
Pang, K-Y. Y. and Miller, K.W. (1978) *Biochem. Biophys. Acta*
  511, 1-9
Purohit, S.C., Bisby, R.H. and Cundall, R.B. (1980) *Int. J.
  Radiat. Biol.* (in press)
Seelig, J. (1978) *J. Prog. Colloid Sci*, 65, 172
Shinitzky, M. (1978) *Biochem. Biophys. Acta* , 515, 367-394

DISCUSSION ON PROF. CUNDALL'S PAPER

Prof. Chapman : What is the PDC concentration in the
membranes of the cells ? Furthermore I notice that you
used a 2:1 ratio with the lecithin liposomes whereas the
cholesterol to lipid ratio in the ghost was different
from this.

Prof. Cundall : We used 0.2 mg/ml PDC and these large
amounts in the membrane may lead to partitioning into the
lipid annulus.

Dr. Johnson : Is urushiol fluorescent ?

Prof. Cundall : Yes, it is slightly fluorescent. It is
however a mixture of a number of components.

Dr. Johnson : As the mixture comes from poison ivy, what
is the component that makes it hypersensitive ?

Dr. Thomas   : This is quite a problem. We don't know
why in such similar compounds one or some are hypersensitive
and others not.

Prof. Porter : What is the cause of the ∿200 nm shift in
the fluorescence of some of the substituted BPH compounds
between hexane and ethanol ?

Prof. Cundall : The compounds are the dimethyl-amino, the
mono-nitro and the methylcyano derivatives. I don't know
of any other compounds that show quite such charge shifts.
The shift depends on macroscopic solvent dielectric
constants and is the same in pure and in mixed solvents.

Prof. Porter : The unsubstituted compound shows only a
minor solvent shift hence one assumes that there are
charge transfer effects causing the spectral shifts i.e.
a dipole is sensitive to solvent polarity.

Prof. Cundall : I agree that charge transfer effects are
possibly the cause but I'm not certain whether one would
expect a linear shift in emission with solvent polarity.

Dr. Teale : What are the fluorescence quantum yields of
these substituted compounds ?

Prof. Cundall : The yields are quite low (<0.1).

Dr. Teale : It seems fairly common that when one gets a large spectral red-shift upon chemical substitution one does not observe a long lifetime compared with the original fluorophor.

Prof. Cundall : The fluorescence lifetimes are indeed short and we hope to use the synchrotron source for future lifetime determinations. As for the chemical lifetime, I don't know how long this is.

Dr. Dale : It is worth remembering that in solution as well as in liposomes, the process of solvent relaxation is time-dependent but in liposomes this is very slow compared to normal solvents, i.e. solvents of low viscosity.

Prof. Cundall : I agree that perhaps in lipid systems we may time-resolve this shift.

Dr. Munro : The DPH on the cells gets into cytoplasm. Do these new derivatives also do so ?

Prof. Cundall : We haven't got that far yet. We simply do not know.

THE USE OF N-(9-ANTHROYLOXY) FATTY ACIDS AS FLUORESCENT PROBES FOR BIOMEMBRANES

KEITH R. THULBORN

Department of Biochemistry
University of Oxford
South Parks Road
Oxford OX1 3QU

INTRODUCTION

A. Membranes

The features of membrane structure pertinent to the following discussion are summarized by the fluid-mosaic model of Singer and Nicolson (1972). Amphiphilic lipid molecules pack together to form a bilayer, with which proteins may interact through electrostatic and hydrophobic forces. A thermodynamic basis for the formation of the bilayer has been described by Israelachvilli, Mitchell and Ninham (1976) in a general theory for the self-assembly of hydrocarbon amphiphiles based on molecular geometries. The effect of proteins on the bilayer has also been considered (Israelachvilli 1977).

Although membrane-bound enzyme, receptor and transport activities are usually associated with the protein components, these functions can be modulated by lipid-protein interactions and by the properties of the lipid matrix (for a review see Gennis and Jonas, 1977). The task of elucidating structure-function relationships in biological membranes is formidable as few techniques are applicable to such complex systems. However as membranes generally lack fluorescent constituents, extrinsic fluorescent probe molecules may be introduced to investigate various aspects of such structures.

B. Fluorescent Probes

In designing fluorescent dyes with which to study membrane

structure, certain criteria must be met.

1. The location of the reporter group must be known and defined
in terms of :
a) the transverse distribution along the normal to the bilayer
surface
b) the lateral distribution across the surface and
c) the exchange of the probe amongst different membranes which
may be present in the system under study.

2. The probe must be sensitive to the membrane properties to
be studied. The parameters to be considered are quantum
yield, fluorescence lifetime and steady-state polarisation.

3. Minimum perturbation should be caused by the incorporation
of the probe.

The extent to which n-(9-anthroyloxy) fatty acids meet these
criteria and may be used to provide some information about
the lipid matrix of membranes will now be described. Emphasis
is given to the limitations of the interpretation of results
and the possible directions that may be taken to extend these
interpretations.

THE N-(9-ANTHROYLOXY) FATTY ACIDS

Figure 1 shows a schematic representation of the set of probes
in which the 9-anthroyloxy reporter group is attached by an
ester linkage to the 2-, 6-, 9-, 12-, and 16- positions of a
long chain fatty acid. The 2- and 16- probes use palmitic
acid and the 6-, 9- and 12- probes use stearic acid. The
methyl ester of 9-anthroic acid was also used for comparison.
12-AS [+] was first used by Waggoner and Stryer (1970), 2-AP

[+] Abbreviations

6-, 9-, 12-AS	:	6-, 9-, 12-(9-anthroyloxy) stearic acid
2-, 16-AP	:	2-, 16-(9-anthroyloxy) palmitic acid
M-9-A	:	methyl-9-anthroate
12-, 16-NS	:	12-, 16-nitroxide stearate
DPH	:	1,6-diphenyl-1,3,5-hexatriene
trans-PA	:	trans parinaric acid
EPR	:	electron paramagnetic resonance
DMPC	:	dimyristoyl phosphatidylcholine
DPPC	:	dipalmitoyl phosphatidylcholine
DSPC	:	distearoyl phosphatidylcholine

was introduced by Barratt *et al.* (1974) and 16-AP was first described by Cadenhead *et al.* (1977). 16-AP will not be discussed in detail here except where relevant comparisons have been made.

The design of such a set of probes assumes that the intrinsic properties of the reporter group do not change with the point of attachment to the acyl chain and that the transverse location of that group is determined by that attachment. If these two assumptions are valid then differences observed in fluorescent properties amongst the probes on incorporation into bilayers can be interpreted directly in terms of variations in membrane environment. By comparing the fluorescent parameters of the probes in organic solvents of different polarities and viscosities, it has been possible to verify that 6-, 9-, and 12-AS have identical spectral characteristics with minor changes being observed for 2-AP, probably due to the close proximity of the fluorophore to the carboxyl group (Thulborn *et al.* 1979). The second assumption is verified below.

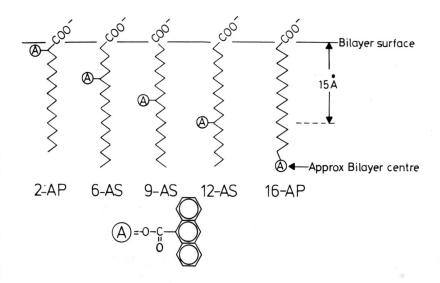

Fig. 1. Schematic representation of the n-(9-anthroyloxy) fatty acids showing the proposed transverse positions of the reporter group, the 9-anthroyloxy moiety, within the bilayer.

A. Location

(a) *Transverse distribution*   The distribution of the reporter
group along the normal to the bilayer surface can be resolved
by fluorescence quenching. Quenching is a short-range interaction requiring the close proximity of the quencher to the
fluorophor. In the case of dynamic quenching in which the
interaction involves the excited state of the fluorophor, the
process is described by the Stern-Volmer relationship :

$$\frac{I_o}{I} - 1 = k_q \tau \{Q\} \qquad (1)$$

where I_o and I are the intensities before and after quenching,
k_q is the bimolecular quenching rate constant, τ is the
fluorescence lifetime before quenching and $\{Q\}$ is the quencher
concentration. Quenching efficiency will thus reflect the
accessibility of the fluorophor to the quencher. If the
location of the quencher is known then the differences in
quenching efficiency amongst the probes must reflect differences in their accessibility and therefore location.
        Cu(III) ions are effective paramagnetic quenchers of
9-anthroyloxy fluorescence. Figure 2 presents the Stern-
Volmer plots for Cu(II) quenching of the probes bound to DMPC
vesicles at 20°C. The order of quenching efficiency, 2-AP >
6-AS > 9-AS > 12-AS > M-9-A, is that expected for a water-
soluble quencher when the location of the reporter group is
governed by its point of attachment to the fatty acid chain.
This however is not the complete explanation as the plots
are clearly nonlinear. Such nonlinearity has been attributed
to the binding of Cu(II) ions to the carboxyl group of the
probes as demonstrated by EPR spectroscopy (Ruzic and
Thulborn, unpublished observations). As the binding sites
become saturated, quenching efficiency decreases resulting
in the downward curvature of the plots. Such binding makes
it difficult to verify that all the probes have the same
intrinsic quenching rate constant k_q in the absence of the
membrane.
        16-Nitroxide stearate (16-NS) quenches 9-anthroyloxy
fluorescence through its nitroxide free radical. Independent
studies have shown that 16-NS positions this radical towards
the centre of the bilayer (Schreier-Muccillo, Marsh and Smith,
1976). Figure 3 shows the Stern-Volmer plots in the same
system as in figure 2. With the exception of 2-AP, the order
of quenching efficiency, M-9-A > 12-AS > 9-AS > 6-AS is
consistent with the reporter groups residing at a graded
series of depths into the bilayer. The anomolously high

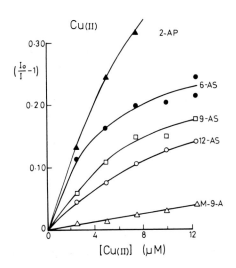

Fig. 2. The Stern-Volmer plots for Cu(II) quenching of the
9-anthroyloxy probes bound to DMPC vesicles (0.5 mM) at 20°C.
(▲) 2-AP, (●) 6-AS, (☐) 9-AS, (o) 12-AS and (△) M-9-A.
Probe to lipid ratio was 1:150 (mole/mole). Cu(II) ions
were added as a CuSO₄ solution. Excitation and emission wavelengths were 365 and 440 nm, respectively. Excitation and
emission slit widths were 6 and 8 nm, respectively.

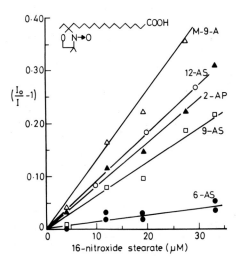

Fig. 3. The Stern-Volmer plots for the quenching of (▲)
2-AP, (●) 6-AS, (☐) 9-AS, (o) 12-AS and (△) M-9-A by
16-NS in DMPC vesicles under the same conditions as in
fig. 2.

quenching for 2-AP can be attributed to the operation of an
additional quenching mechanism as indicated by the different
spectral changes. This may result from changes in surface
charge due to increasing 16-NS concentration.
    The indole moiety of N-steroyl tryptophan can be shown to
be located in the polar headgroup region of the bilayer by
comparison of its emission spectrum in bilayers to those in
organic solvents of different polarities. Figure 4 presents
the Stern-Volmer plots for the quenching of the probes in
DMPC multilayers at 20°C by such a group. The order of
quenching efficiency, 2-AP > 6-AS > 9-AS ∿ 12-AS > M-9-A is
consistent with the location of the reporter groups predicted
on the basis of the probe design.

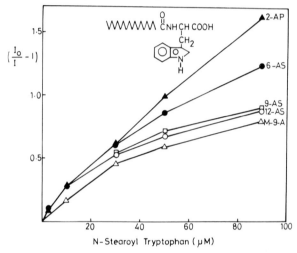

N-Stearoyl Tryptophan (µM)

*Fig. 4. The Stern-Volmer plots for the quenching of (▲)
2-AP, (●) 6-AS, (◻) 9-AS, (o) 12-AS and (△) M-9-A by
N-stearoyl tryptophan in DMPC multilayers (0.5 mM) at 22°C.
Instrument settings and probe to lipid ratio as in fig. 2.*

    The use of a number of different quenching molecules has
established that the probes do position their reporter
groups at depths into the bilayer dependent on the point of
attachment to the fatty acid chain.

(b) Lateral distribution   The distribution of the probes
across the surface of a bilayer will be determined by the
ease with which these bulky molecules can be accommodated
within the closely packed acyl chains of the phospholipids.
The more disordered regions may allow the probes to be more
readily incorporated. Membranes in which lateral segregation
of lipids may occur, such that different lipids exist in

segregated domains of different fluidity, may show non-uniform distributions of probe molecules. Their fluorescent parameters would then yield information only about the host domains.

Lateral phase separation can be simulated by a binary mixture of DMPC and DSPC in multilayers (De Kruijff *et al.*, 1974). Figure 5 displays the steady-state polarisation, a measure of the extent of fluorophor rotation during its excited lifetime, as a function of temperature for 12-AS bound to DMPC, DSPC and DMPC:DSPC (1:1, mole/mole) multi-layers. The same results were obtained for 2-AP, 6-AS and 9-AS. Light scattering for the binary mixture is also plotted.

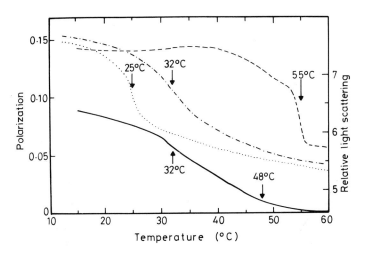

Fig. 5. Polarisation of 12-AS bound to (·····) DMPC, (---) DSPC and (·—·—·—) DMPC:DSPC (1:1, mole/mole) multilayers (0.5 mM) as a function of temperature. Light scattering at 365 nm (——) as a function of temperature is also shown for the binary mixture. Heating rate was 1°C/min. Probe to lipid ratio was 1:150 (mole/mole). The fluorescence polar-isation instrument described by Thulborn and Sawyer, 1978 gives a continuous readout of polarisation and intensity and of scattering light after removal of the emission cutoff filters.

A decrease in polarisation reflects increased probe rotation in the simplest interpretation of this parameter (see below). The sharp changes seen at 25°C and 55°C for the DMPC and DSPC multilayers, respectively, are in good agreement with

the gel to liquid-crystalline transitions reported by
differential scanning calorimetry. For the binary mixture,
two transitions at 32°C and 48°C are observed by light
scattering corresponding to the melting of the two different
phospholipid domains of DMPC and DSPC, respectively. The
values are shifted towards intermediate values from those of
pure lipids due to the incomplete separation of the lipids
within the mixture (De Kruijff et al. 1974). In contrast,
12-AS reports on only the lower transition at 32°C, indicating
that these probes are located preferentially in the most
fluid domains of the membranes.

Such behaviour of the n-(9-anthroyloxy) fatty acids is
contrasted to that of other fluorescent probes, 1,6-diphenyl-
1,3,5-hexatriene (DPH) and trans-parinaric acid, in figure
6 in the same binary mixture as above. Whereas 12-AS detects
only the lower transition, trans-PA monitors only the upper
transition and therefore prefers the gel state (Sklar, Hudson
and Simoni, 1977). DPH reports one broad transition over the
full range of the two phase changes which is consistent with
the partitition coefficient of unity between fluid and gel
phases measured by Lentz, Barenholz and Thomson, 1976.

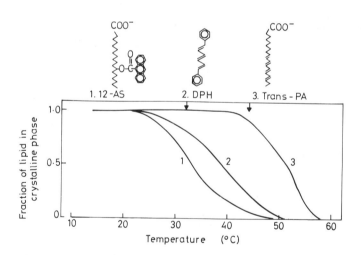

*Fig. 6. Comparison of the thermotropic phase transitions
monitored by the steady-state polarisation of (1) 12-AS,
(2) DPH and (3) trans-PA in DMPC:DSPC (1:1 mole/mole)
multilayers (0.5 mM). All polarisations were normalised
to unity at 20°C. Conditions as in fig. 5.*

It is clear that the lateral distribution of fluorescent
probes need not be uniform and that the n-(9-anthroyloxy)

fatty acids are an example of probes favouring the more fluid
regions of phosphatidylcholine bilayers. Other types of
phospholipid have yet to be examined.

(c) Exchange   The exchange of probes between membranes will
not be discussed except to mention that it is slow for the
fatty acid probes but very rapid for M-9-A. It would appear
that the fatty acid chain is a good anchor for fluorescent
dyes in membranes.
    Having shown that the location of these probes can be
defined and therefore that the first criterion is satisfied,
the interpretation of some of the fluorescence parameters
will now be described.

B. Polarity

Frequently the fluorescence quantum yield and lifetime of
probe molecules are sensitive to both the viscosity and the
polarity of their environment. However the n-(9-anthroyloxy)
fatty acids show great sensitivity to the polarity of organic
solvents and paraffin oils but not to the viscosities. These
parameters can thus be used to monitor transverse polarity
gradients into the bilayer.
    Figure 7 presents the plots of fluorescence lifetime and
relative quantum yield as a function of the attachment
position carbon number for the probes in DMPC vesicles at 20°C.

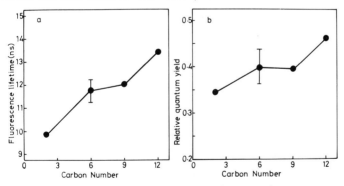

*Fig. 7.   Fluorescence lifetimes and relative quantum yields
of the n-(9-anthroyloxy) fatty acids as a function of n (=
2, 6, 9, 12) in DMPC multilayers at 20°C. Concentrations and
settings as in fig. 2. Quantum yields relative to a quinine
sulphate standard. Lifetimes measured by single photon
counting using 363 nm excitation and a combination of 400
and 420 nm cut-off filters on the emission side to minimise
scattering. Error bars are representative of all measurements.*

The lifetimes were deconvoluted as single exponential decays.
(Subsequent measurements have shown the decays to be
biexponential, Thulborn and Beddard, unpublished results).
Both parameters show a gradual increase towards the centre of
the bilayer consistent with decreasing polarity and hence
decreasing water penetration on the nanosecond time scale.
The consistency of both parameters determined by independent
methods - although both are related by the radiative rate
constant - lends confidence to such an interpretation. The
question of whether the polarity gradient observed in the
presence of the probes is the same in their absence is
difficult to answer. The problem of perturbation is discussed
below.

C. Fluidity

Steady-state fluorescence polarisation is reported extensively
in the literature as a means of monitoring membrane fluidity
(Shinitzky and Inbar, 1976;Lentz et $al.$ 1976). The
n-(9-anthroyloxy) probes provide a means of measuring a
fluidity gradient into the bilayer if it can be shown that
they all monitor the same polarisation in the same environment. Figure 8 shows the Perrin plots for the probes
dissolved in glycerol over a wide range of viscosities. The
fatty acid probes show identical behaviour whereas M-9-A
records lower polarisations. Its smaller molar volume
apparently allows it greater rotational freedom that the other
probes and hence its polarisation cannot be compared directly
to those probes. 16-AP has also been demonstrated to record
lower polarisation values due to the greater rotational
freedom of the terminal methylene involved in its ester
linkage (Tilley, Thulborn and Sawyer, 1979).
    The polarisation gradients in DPPC multilayers in the gel
state at 20°C and in the fluid state at 47°C are recorded
in figure 9. The shape of the gradient is consistent with
EPR results using nitroxide-labelled fatty acids (Hubbell
and McConnell, 1971) with decreasing polarisation indicating
greater fluidity towards the centre of the bilayer. It must
be noted that fluidity in this sense refers to the ability of
a reporter group to rotate at a particular location within
the hydrocarbon region of the bilayer and does not represent
only the motional freedom of the phospholipid acyl chains as
reported using deuterium nmr (Seelig and Seelig, 1974).
    The probes can also be used to monitor events as a function
of depth during phase transitions as indicated in figure 10
in which polarisation is plotted against temperature for each
probe bound to DPPC multilayers. The main transition at
41.6°C is clearly seen by all the probes, with the size of

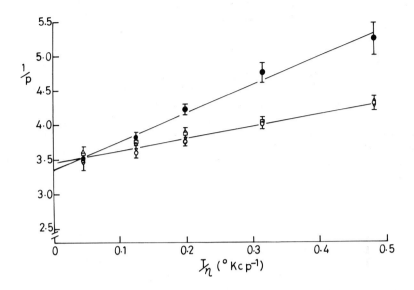

Fig. 8.   Perrin plots for 2-AP (▲), 6-AS (△), 9-AS (□), 12-AS (○) and M-9-A (●) dissolved in glycerol.   Error bars show the scatter of values from triplicate measurements at probe concentrations of 50 µM.   Excitation and emission wavelengths were 365 and 440 nm, respectively.   Excitation and emission slit widths were 6 and 30 nm, respectively.

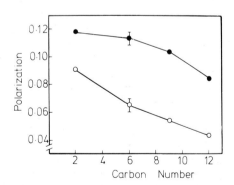

Fig. 9.   Polarisation of the n-(9-anthroyloxy) fatty acids as a function of n (n = 2, 6, 9, 12) for the probes bound to DPPC vesicles (0.5 mM) at 47°C (○) and 20°C (●). Error bars are representative of all points.   Probe to lipid ratio was 1:150 (mole/mole).

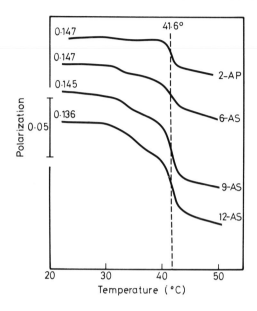

Fig. 10. Polarisation of the 9-anthroyloxy probes as a function of temperature in DPPC multilayers (0.5 mM). Measurements were made as described in fig. 5.

the pretransition at 35^OC increasing from 2-AP to 12-AS, that is, towards the centre of the bilayer. The pretransition is a feature of thermotropic transitions in multilayers, a preparation formed by vortexing phospholipids in aqueous solution. The smaller single-walled vesicles formed by sonification show no such feature. Even in multilayers, the pretransition is seen only on heating, never on cooling at even very slow cooling rates, in contrast to the freely reversible main transition. The above results are consistent with x-ray scattering studies which have ascribed the pre-transition to the reorientation of the phospholipid acyl chains (Janiak *et al.*, 1976; Brady and Fein, 1977).

The narrow main transition registered by all the probes is in excellent agreement with differential scanning calorimetry and indicates that perturbation by the probes was not significant in these experiments.

Having used an oversimplified interpretation of polaris-ation in simple lipid systems, it is now necessary to consider more complex models and eventually whole cells. The incorp-oration of cholesterol is the next stage of complexity. Figure 11 compares the phase transitions measured by the polarisation of 12-AS and DPH for DPPC and DPPC-cholesterol (45 mole%) multilayers. In both cases cholesterol broadens

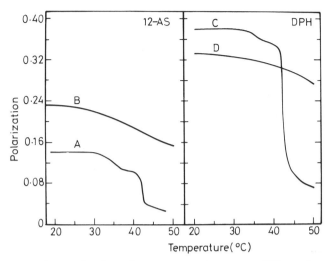

Fig. 11. The polarisation-temperature profiles for 12-AS in DPPC multilayers (A) and DPPC-cholesterol (1:1, mole/ mole) multilayers (B) and DPH in DPPC multilayers (C) and DPPC-cholesterol (1:1, mole/mole) multilayers (D). Measurements were made as described in fig. 5.

the transitions as reported by other techniques (Ladbrooke and Chapman, 1969; Lippert and Peticolas, 1971). However an important difference is that 12-AS reports higher polarisations at all temperatures in the presence of cholesterol whereas DPH shows higher polarisations at temperatures above the main transition but lower values below that transition in the presence of cholesterol.

It is well known from a variety of techniques that the rigid steroid ring structure of cholesterol makes the phospholipid acyl chains of the formerly fluid state less mobile but disrupts the ordered packing of the gel state (Lippert and Peticolas, 1971; Schreier-Muccillo *et al.*,1973). The question now arises as to why DPH monitors the correct cholesterol effects but 12-AS does not.

The explanation is based on the different orientations of the emission oscillators of the probes in the bilayer as indicated in figure 12. DPH has its emission dipole perpendicular to the surface along the long axis of the molecule (Andrich and Vanderkooi, 1976; Cehelnik *et al.*, 1973). In contrast, 12-AS has its emission oscillator only 30⁰ away from being parallel to the bilayer surface (Badley *et al.*, 1973). Changes in polarisation are induced only by motions that change the direction of these emission oscillators. Such changes may involve either a change in the rate of probe motion in which case the fluidity is changing or a change in the modes of rotation which reflects an altered anisotropy of probe motion. Steady-state polarisation thus has contributions from both the rate and the anisotropy and one must consider which factor is dominant in a particular situation.

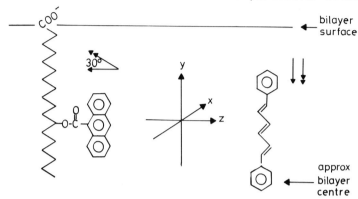

Fig. 12. The orientation of 12-AS (left) and DPH (right) in the lipid bilayer. The single and double-headed arrows indicate the directions of the absorption and emission transition moments, respectively. The diagram is based on the work of Badley et al., 1973 on 12-AS and of Andrich and Vanderkooi, 1976 and Cehelnik et al., 1973 on DPH.

Axial rotation around the long axis of DPH does not change the direction of its emission oscillator and this leads to no resultant change in polarisation. The same rotation for 12-AS is however very effective in changing polarisation. It seems reasonable that the large aromatic ring structures of the probes may pack parallel to the plane of the steroid ring structure to mimimise packing energies within the bilayer. Such packing would limit axial rotations and thus explain the different behaviours of the two probes displayed in figure 11.

The influence of cholesterol on the anisotropy of the n-(9-anthroyloxy) fatty acids is emphasized in figure 13 in which the polarisation gradients in human erythrocyte ghosts and in rat hepatocytes are presented. The salient feature is the much higher polarisation values for the ghost preparation. Although it may have been concluded that hepatocytes have more fluid membranes, in fact the difference is largely a reflection of the very high cholesterol content of the ghosts. The difficulty is resolving the two contributions to steady-state polarisation should be largely overcome by the use of time-resolved emission anisotropy measurements (Kinosita, Kawato and Ikegami, 1977).

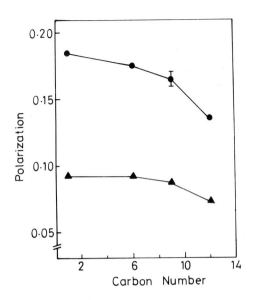

Fig. 13. Polarisation gradients in human erythrocyte ghosts (●) and in rat hepatocytes (▲) at 20°C. Probe concentrations were 3.3 μM. The error bar is representative of errors for all points.

PERTURBATION

The most serious crticism levelled at the use of extrinsic probe molecules is that of perturbation caused by the probe to the environment that it is monitoring. Such disturbances have been reported at high probe to lipid ratios. Changes in the X-ray diffraction electron density profiles of bilayers containing 20 mole% 12-AS have been reported (Lesslauer, Cain and Blaise, 1972). Podo and Blaise (1977) demonstrated changes in the nmr longitudinal relaxation times of protons in the acyl chains of phospholipids in bilayers doped with 25 mole% of 2-AP and of 12-AS. Cadenhead *et al.*, (1977) reported that differential scanning calorimetry showed depressed phase transition temperatures for bilayers containing 20 mole% of 2-AP and of 12-AS. Such perturbations at these concentrations is not unexpected.

Few attempts have been made to determine the extent to which disruption may be important under the conditions of the fluorescence experiment, that is, at fluorophor concentrations of less than 1 mole%.

As has been described above, the n-(9-anthroyloxy) fatty acids do monitor the same phase transition temperature as non-perturbing techniques (figures 5, 10). No change in

bilayer permeability to glucose could be detected on incorporation of the probes (Bui, 1978). In contrast, electrical
measurements on bimolecular lipid membranes show changes down
to probe concentrations as low as 0.01 mole% 2-AP (Ashcroft
et al., in print). Such changes have been interpreted in
terms of alterations in bilayer packing and thickness. 2-AP
has been observed to cause concentration dependent lysis of
human erythrocytes and lymphocytes that is not observed for
the other derivatives (Bui, 1978; Howard and Sawyer, 1979).

It would seem that there are situations in which specific
interactions and perturbations do preclude the use of these
probes. However to exclude the use of the probes under all
circumstances would be to remove one of the few techniques
which can provide some information, albeit only a first
approximation, about such complex systems as natural membranes.

FLUORESCENCE QUENCHING

Having described a simple interpretation of quenching in
bilayers in order to establish the transverse distribution of
the n-(9-anthroyloxy) fatty acids, a more general treatment
for studying a variety of membrane interactions is now
presented.

The Stern-Volmer equation (eq. 1) strictly applies to
fluorescence quenching in a single-phase system. An aqueous
phospholipid dispersion represents a two-phase system and so
the equation must be recast into the three models shown in
figure 14. Model 1 describes the distribution of quencher
between the two phases by a partition coefficient K_p and
assumes that quenching only occurs within the lipid phase.
Model 2 treats the quencher as binding directly to the
fluorophor, accessible at the interface of the two phases,
using an equilibrium association constant K_A. Model 3 assumes
that the quencher binds to an acceptor within the bilayer,
as described by an equilibrium association constant K_B. The
bound quencher may then diffuse throughout the lipid phase
to quench the fluorophor. The derivations of the equations
required to obtain the partition coefficient or association
constants together with bimolecular quenching rate constants
are given in Appendix 1. The quenching rate constant provides
a simple method of calculating diffusion coefficients within
bilayers.

Figure 15 presents the Stern-Volmer plots for the quenching
of 12-AS bound to three different concentrations of DMPC
vesicles using 12-AS as quencher. The curves through the
experimental points are calculated from the data analysis of
model 1 as described in Appendix 1. The upward curvature is
due to a static contribution to the quenching process.

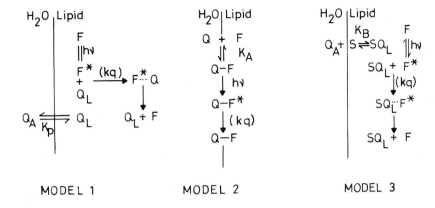

MODEL 1   MODEL 2   MODEL 3

*Fig. 14.   Three models for fluorescence quenching of a
fluorophor in a two-phase system.   Model 1 describes the
quencher distribution by a partition coefficient (K_p); model
2 assumes that the fluorophor is freely accessible to the
quencher in the aqueous phase and describes the binding by
an equilibrium association constant (K_A) and model 3 describes
the binding of quencher to the same phase as the fluorophor
by an equilibrium association constant (K_B).   The bimolecular
rate constant is k_q.*

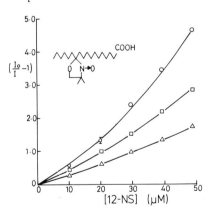

*Fig. 15.   The Stern-Volmer plots for the quenching of 12-AS
by 12-NS in DMPC vesicles at 20°C.   Phospholipid concentrations
were 0.6 mM (△), 0.4 mM (□) and 0.2 mM (○).   Computed
curves through the experimental points are based on model 1
with parameter estimates of W =3.9 ± 0.8 M^{-1}, K_p= 5 x 10^{-3},
K_q =1.5 x 10^9 $M^{-1}s^{-1}$.   Probe to lipid ratio was constant
at 1:400 (mole/mole).*

Figure 16 shows the Stern-Volmer plots for Cu(II) ion quenching of 2-AP bound to DMPC vesicles at two different 2-AP concentrations. The curves through the experimental points were computed from the constants derived from model 2 assuming a 1:1 stoichiometry.

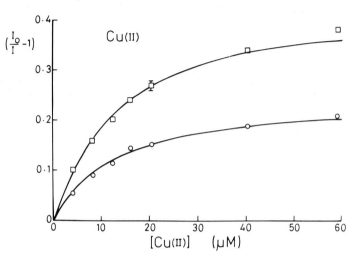

Fig. 16. The Stern-Volmer plots for Cu(II) quenching of 2-AP in DPPC vesicles at 20°C for 1.7 μM 2-AP (o) and 5 μM 2-AP (□) in 0.5 mM phospholipid. Computed curves through the experimental points are based on model 2 with parameter estimates of $K_A = 1 \times 10^5 \; M^{-1}$ and $k_q = 1 \times 10^{13} \; M^{-1} \, s^{-1}$.

The computed fits to the experimental data based on models 1 and 2 and the agreement with the literature values for the partition coefficient of 12-NS (Butler, Tattrie and Smith, 1974) and for the binding constant of divalent cations to carboxyl groups (Hauser, Darke and Phillips, 1976; Puskin, 1977) gives support to the use of these analyses.

Another strategy in quenching experiments is to use the known location of the probes to find the location of the quenching species. An example of where this approach has been useful is in determining the orientation of gramicidin-A in bilayers (Haigh *et al.*, 1979). This represents an example of model 3. However as the stoichiometry of gramicidin-A binding to bilayers is not known, analysis has not yet been attempted.

Gramicidin-A is a linear polypeptide antibiotic of 15 amino acids in which positions 9, 11, 13 and 15 are L-tryptophan. The antibiotic renders lipid bilayers permeable to alkali cations and protons by forming a transbilayer ion channel involving two gramicidin molecules. Urry (1971)

proposed that the channel formed by a 'head to head' dimer-
ization involving the interaction of the blocked N-terminal
ends. Such a model places the tryptophan residues at the
surface. Other workers (Veatch *et al.*, 1974) have suggested
that parallel and antiparallel β double-helical dimers may
form which would distribute the tryptophans throughout the
bilayer.

The quenching of the fatty acid probes by gramicidin-A
in DMPC liposomes at 20°C is shown as Stern-Volmer plots in
figure 17. The order of quenching efficiency, 2-AP > 6-AS >
9-AS > 12-AS > 16-AP, indicates that the indole moities must
be closest to the 2-AP fluorophor at the glycerol backbone.
Such a result is consistent with the dimer model of Urry (1971).

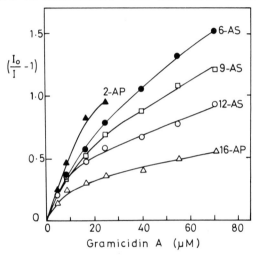

*Fig. 17. The Stern-Volmer plots for gramicidin quenching of
2-AP (▲), 6-AS (●), 9-AS (□), 12-AS (O) and 16-AP
(Δ) bound to DMPC multilayers (0.5 mM) at 20°C. Probe to
phospholipid ratio 1:100 (mole/mole). Curves are drawn by
hand.*

CONCLUSIONS

The n-(9-anthroyloxy) fatty acids largely satisfy the three
criteria initially proposed for using extrinsic fluorescent
probes in membranes. They can obtain information not readily
accessible by other techniques at this time. However care
must be taken not to oversimplify the analysis of fluorescence
data. Steady-state polarisation has clear limitations for
these probes and only more sophisticated time-resolved emission
anisotropy measurements can provide a thorough interpretation.
Fluorescence quenching analyses as described promise to be

useful in quantitative studies of a variety of membrane
interactions. There remains the possibility of artefacts
arising from perturbation by the probes and this must be
considered in the light of each new experiment.

ACKNOWLEDGEMENTS

The work described above was presented as a Ph.D. thesis at
the University of Melbourne. The author gratefully acknowledges the enthusiastic supervision of Dr. W.H. Sawyer over
the course of that thesis. The author is currently a recip-
ient of a Commonwealth Scientific and Industrial Research
Organization post-doctoral research studentship.

APPENDIX 1

Theory for quenching in two-phase systems

The dynamic quenching of a fluorophor in its excited state
in free solution is described by the Stern-Volmer equation.
If static quenching also occurs, involving the formation of
a non-fluorescent ground-state fluorophor-quencher complex,
then according to Weller (1961), equation 1 becomes :-

$$\frac{I_0}{I \exp(W\{Q\})} = 1 + k_q \tau \{Q\} \tag{2}$$

where W is the static quenching constant and other terms have
been previously defined. The term $\exp(W \{Q\})$ can be simplified
to $(1 + W \{Q\})$ if 1:1 stoichiometry for the ground-state
complex is assumed (Moon, Poland and Scheraga, 1965). It must
be remembered that intensities (I, I_0) can only be used
instead of quantum yields if there is no change in the shape
of the emission spectrum during quenching.
   If both dynamic and static quenching processes occur, then
the Stern-Volmer plot is non-linear and shows upward curvature
from the plot of dynamic quenching alone. The two contributions to the overall process can be resolved by plotting
$I_0/I \exp(W'\{Q\})$ vs. $\{Q\}$ for varying values of W' until a
linear plot is obtained (Eftink and Ghiron, 1976). k_q is
then determined from the slope of this plot. Alternatively
a direct determination of the dynamic contribution can be
made by the measurement of fluorescence lifetimes and a plot
of τ_0/τ vs. $\{Q\}$ (Chance, Erecinska and Radda, 1975).
   If a fluorophor in a phospholipid dispersion is bound
exclusively to the bilayer phase or is fluorescent only in
that phase due to the polarity dependence of quantum yield,

then quenching will be dependent on the effective concentration of quencher in the lipid phase. Three models for
describing this effective concentration are presented in
figure 14. The equations are now developed.

(a) Model 1 : Partition

The distribution of a quancher Q between the aqueous (subscript A) and lipid (subscript L) phases may be described by
a partition coefficient :

$$K_p = \{Q\}_L / \{Q\}_A \tag{3}$$

where $\{Q\}_L$ and $\{Q\}_A$ refer to the concentrations of $\{Q\}$
expressed in terms of the volumes of each phase, V_L and V_A
respectively. Using $V_T \{Q\} = V_A \{Q\}_A + V_L \{Q\}_L$ where
Q is the concentration with respect to the total sample
volume, $V_T = V_A + V_L$, then equation (1) becomes :

$$\frac{I_o}{I} - 1 = \frac{k_q \tau K_p V_T \{Q\}}{V_A + V_L K_p} \tag{4}$$

The Stern-Volmer plot using the total quencher concentration
is thus linear over the range in which the partition coeff-
cient remains constant with a slope S given by :

$$S = \frac{k_q \tau K_p V_T}{V_A + V_L K_p} \tag{5}$$

It is not possible to determine K_p and k_q from a single
Stern-Volmer plot. However by rearrangement of equation (4)
and as $V_L / V_A \ll 1$ in fluorescence experiments then :

$$\frac{I}{I_o - 1} \{Q\} = \frac{1}{k_q \tau K_p} + \frac{1}{k_q \tau} \left(\frac{V_L}{V_A}\right) \tag{6}$$

By plotting $I \{Q\} /(I_0 - 1)$ $vs.$ (V_L/V_A), values of K_p and
k_q can be obtained from the slope and intercept, assuming
that they are independent of the total lipid concentration.
Such an assumption is reasonable at quencher and fluorophor
concentrations well below the saturation level of the lipid
phase.
    In cases of quenching with both dynamic and static
contributions, equation (2) must be used and becomes :

$$\frac{I_o}{I \exp(W'\{Q\})} = 1 + \frac{k_q \tau \, K_p \, V_T \, \{Q\}}{V_A + V_L K_p} \qquad (7)$$

where the apparent static quenching constant W' is related to W by :

$$W' = \frac{W \, K_p \, V_T}{V_A + V_L \, K_p} \qquad (8)$$

By changing W' until a plot of $I_o/I\exp(W' \{Q\})$ *vs.* $\{Q\}$ is linear. the static contribution can be eliminated. Analogous equations to equation (6) can then be derived to obtain K_p and k_q.

An alternative procedure to using equation (6) is to use the ratio of slopes (S and S') obtained from two Stern-Volmer plots (corrected for the static contribution if necessary) at the same total volume V_T but for different lipid concentrations (C and C', respectively). Thus, using equation (5) and a molar volume of phospholipid of V_m, then as $V_T = V_A + V_L = V_A' + V_L'$:

$$\frac{S}{S'} = \frac{V_T + C' \, V_m(K_p - 1)}{V_T + C \, V_m \, (K_p - 1)} \qquad (9)$$

K_p can then be determined as the only unknown. Resubstitution of K_p into equation (5) allows k_q to be found.

(b) Model 2 : Binding to the fluorophor

The quenching of a fluorophor which binds strongly to a quencher will be dominated by the binding process. For large association constants and low acceptor concentrations, only the bound quencher need be considered as contributing to quenching.

The equilibrium for the interaction can be written as :

$$Q + nF \rightleftarrows F_n Q$$

where n is the number of fluorophors F, binding to each quencher. If α is the mole fraction of free quencher at equilibrium, then the equilibrium association constant can be given as :

$$K_A = \frac{1 - \alpha}{\alpha \{F\}^n} \qquad (10)$$

For a two-phase system in which the quencher and fluorophor are in different phases, their interaction must occur at the phase boundary. Equation (1) can now be written as :

$$\frac{I_o}{I} - 1 = k_q \tau\{F_n Q\} = k_q \tau (1 - \alpha)\{Q\} \tag{11}$$

which by rearrangement of (10) and substitution into (11) gives :

$$\frac{I_o}{I} - 1 = \frac{k_q \tau\{Q\}}{1 + \dfrac{1}{(K_A\{F\}^n)}} \tag{12}$$

where $\{F\}$ is the equilibrium concentration of free fluorophor. This assumes that each site on the quencher causes an equivalent amount of quenching.

The determination of k_q and K_A from such Stern-Volmer plots is now considered. In the limit as $\{Q\}$ approaches zero, $\{F\}$ approaches $\{F\}_T$, a result readily demonstrated analytically and the limiting slope, S_i is :

$$S_i = \frac{k_q}{1 + \dfrac{1}{(K_A \{F\}_T^n)}} \tag{13}$$

Assuming that k_q and K_A are independent of $\{F\}_T$, the total fluorophor concentration, the ratio of the limiting slopes (S_i and S_i') of two Stern-Volmer plots made at two different fluorophor concentrations ($\{F\}_T$ and $\{F\}_{T'}$, respectively) is

$$\frac{S_i}{S_i'} = \frac{1 + 1/K_A \{F\}_{T'}^n}{1 + 1/K_A \{F\}_T^n} = r \tag{14}$$

which simplifies for $n = 1$ to :

$$K_A = \frac{r\{F\}_{T'} - \{F\}_T}{\{F\}_T\{F\}_{T'} (1 - r)} \tag{15}$$

Given a stoichiometry of 1:1, K_A can be found and, by substitution into (13), k_q can be determined. As limiting slopes are difficult to measure accurately, the above procedure should be considered as a means of estimating K_A and k_q for a curve fitting procedure based on equatuon (12) in which the experimental Stern-Volmer plots are fitted over the entire concentration range used.

An alternative treatment of binding is the classical analysis of Klotz (1946). This would be required in the cases where the stoichiometry is unknown.

(c) Model 3 : Binding to the Lipid Phase

This model assumes that the quencher has a high affinity for the same phase as the fluorophor but that its binding occurs to freely accessible sites (S) which are topographically distinct from those of the fluorophor in the bilayer. Diffusion of the quencher and fluorophor within the bilayer then determines the rate of quenching. The binding equilibrium is :

$$S_{(lipid)} + Q_{(aq)} \rightleftharpoons SQ_{(lipid)}$$

where S is the acceptor with one binding site. Within the bilayer, S may be made up of a number of phospholipid molecules surrounding the quencher. If α is the mole fraction of free quencher at equilibrium then the association constant K_B is give by :

$$K_B = \frac{\{SQ\}}{\{S\}\{Q\}} = \frac{1 - \alpha}{\alpha \, S} \qquad (16)$$

Equation (1) must be re-expressed with respect to the volume of the lipid phase in which the quenching occurs thus :

$$\frac{I_o}{I} - 1 = k_q \, \tau \, \{SQ\}_L \qquad (17)$$

where $\{SQ\}_L$ is the bound quencher concentration in terms of the volume of the lipid phase. This can be re-expressed in terms of $\{SQ\}$, the bound quencher concentration with respect to the total sample volume V_T for a given lipid concentration C and lipid molar volume V_m, as :

$$\frac{I_o}{I} - 1 = \frac{k_q}{C\,V_m}\,\{SQ\}$$

$$= \frac{k_q\,\tau(1 - \alpha)\,\{Q\}}{C\,V_m} \tag{18}$$

Rearrangement of (16) and substitution into (18) gives :

$$\frac{I_o}{I} - 1 = \frac{k_q\,\tau}{C\,V_m}\,\frac{\{Q\}}{1 + 1/(K_B\,\{S\})} \tag{19}$$

This equation is analogous to (12) in model 2 except that the quenching efficiency is dependent on lipid concentration rather than fluorophor concentration.

Assuming K_B and k_q are independent of C, then the ratio of of the limiting slopes (S_i and S_i') of two Stern-Volmer plots at two different lipid concentrations (C and C', respectively) is given as :

$$\frac{S_i}{pS_i} = \frac{1 + 1/(pK_B\,\{S\}_T)}{1 + 1/(K_B\,\{S\}_T)} \tag{20}$$

where $p = C'/C$ and $\{S\}_T' = p\{S\}_T$, that is, the number of binding sites is proportional to the number of lipid molecules. This equation allows the product $K_B\{S\}_T$ to be calculated which on resubstitution into (19) allows k_q to be found. The determination of K_B requires the use of the binding function analysis of Klotz (1946).

In all three models it has been tacitly assumed that all the fluorophors are potentially accessible for quenching. This need not be the case and the modified Stern-Volmer relationship in which a proportion of the fluorophors are inaccessible to quencher in one phase systems (Lehrer, 1971) should replace equation (1) in the derivatives above. These variations will be discussed elsewhere.

138   THULBORN

REFERENCES

Andrich, M.P. and Vanderkooi, J.M. (1976) *Biochemistry* 15, 1257-1261

Ashcroft, R.G., Thulborn K.R., Smith, J.R., Coster, H.G.L.and Sawyer, W.H.,(1980) *Biochim. Biophys. Acta*, 602, 299-308

Badley, R.A., Martin, W.G. and Schneider, H. (1973) *Biochemistry*, 12, 268-275

Barratt, M.D., Badley, R.A., Leslie, R.B., Morgan C.G. and Radda, G.K. (1974) *Eur. J. Biochem.* 48, 595-601.

Brady, G.W. and Fein, D.B. (1977) *Biochim. Biophys. Acta* 464, 249-259

Buï Hung-Hoon, A. (1978) Masters Thesis, University of Melbourne

Bulter, K.W., Tattrie N.H. and Smith, I.C.P. (1974) *Biochim. Biophys. Acta* 363, 351-360

Cadenhead, D.A., Kellner, B.M.J., Jacobson, K. and Papahadjopoulos, D., (1977) *Biochemistry* 16, 5386-5392.

Cehelnik, E.D., Cundall, R.B., Timmons, C.J. and Bowley, R.M. (1973) *Proc. Roy. Soc. Lond.* A335, 387-405

Chance, B., Erecinska, M. and Radda, G.K., (1975) *Eur. J. Biochem.* 54, 521-529

De Kruijff, B., Van Dijck, P.W.M., Demel, R.A., Schuijff, A., Brants, F and Van Deenan, L.L.M. (1974)*Biochim. Biophys. Acta* 356, 1-7

Eftink, M.R. and Ghiron, C.A., (1976) *J. Phys. Chem.* 80 486-493

Gennis, R.B. and Jonas, A., (1977)*Ann. Rev. Biophys. Bioeng.* 6, 195-238.

Haïgh, E.A., Thulborn, K.R. and Sawyer, W.H., (1979) *Biochemistry* 18, 3525-3532

Hauser, H., Darke, A., and Phillips, M.C. (1976) *Eur. J. Biophys.* 62, 335-344

Howard, R., and Sawyer, W.H. (1981)*Parasitology (in press)*

Hubbell, W.L. and McConnell, H.M. (1971) *J. Am. Chem. Soc.* 93, 314-326

Israelachvilli, J.N. (1977) *Biochim. Biophys. Acta* 469, 221-225

Israelachvilli, J.N. Mitchell, D.J. and Ninham, B.W. (1976) *J. Chem. Soc. Faraday Trans.*, II, 72, 1525-1567

Janiak, M.J., Small, D.M. and Shipley, G.G. (1976) *Biochemistry* 15, 4575-4580

Kinosita, K., Kawato, S., and Ikegami, A. (1977) *Biophys. J.* 20, 289-305

Klotz, I.M. (1946) *Arch. Biochem.*, 9, 109- 117

Ladbrooke, B.D. and Chapman, D. (1969) *Chem. Phys. Lipids* 3, 304-367

Lehrer, S.S. (1971)*Biochemistry*   10, 3255-3263
Lentz, B.R., Barenholz, Y. and Thompson, T.E. (1976)
  Biochemistry 15, 4521-4528
Lesslauer, W., Cain, J.E., and Blaise, J.K.(1972) *Proc. Nat.*
  Acad. Sci. U.S.A. 69, 1499-1503
Lippert, J.L. and Peticolas, W.L. (1971) *Proc. Nat. Acad.*
  Sci. U.S.A. 68, 1572-1576
Moon, A.Y., Poland, D.C. and Scheraga, H.A. (1965) *J. Phys.*
  Chem. 69, 2960-2966
Podo, F. and Blaise, J.K. (1977) *Proc. Nat. Acad. Sci.*
  U.S.A. 74, 1032-1036
Puskin, J.S. (1977) *J. Membrane Biol.* 35, 39-55
Schreier-Muccillo, S., Marsh, D., Dugas, H., Schneider H.
  and Smith, I.C.P. (1973) *Arch. Biochem. Biophys.*
  172, 1-11
Seelig A. and Seelig, J. (1974) *Biochemistry* 13, 4839-4845
Shinitzky, M. and Inbar, M. (1976) *Biochim. Biophys. Acta*
  433, 133-149
Singer, S.J. and Nicolson, G.L. (1972) *Science* 175, 720-731
Sklar, L.A., Hudson, B.S. and Simoni, R.D. (1977)
  Biochemistry 16, 819-828
Thulborn, K.R. and Sawyer, W.H.,(1978) *Biochim. Biophys. Acta*
  511, 125-140
Thulborn, K.R., Tilley, L.M., Sawyer, W.H. and Treloar, F.E.
  (1979)*Biochim. Biophys. Acta* 558 , 166-178
Tilley, L.M., Thulborn, K.R., and Sawyer, W.H. (1979)
  J.Biol. Chem., 254, 2592-4
Urry, D.W. (1971) *Proc. Nat. Acad. Sci. U.S.A.*
  68, 672-676
Veatch, W., Fossel, E.T. and Blout, E.K. (1974) *Biochemistry*
  13, 5249-5256
Waggoner, A.S. and Stryer, L. (1970) *Proc. Nat. Acad. Sci.*
  U.S.A. 67, 579-589
Weller, A. (1961) *Prog. React. Kinet.* 1, 187.

DISCUSSION ON DR. THULBORN'S PAPER

Dr. Dale : Why does the DPH depolarisation not show the
presence of two phase transitions in the binary phospholipid mixture ?

Dr. Thulborn : The reason for that is that we are using
polarisation and anisotropy. Polarisations do not necessarily add so I don't think that the interpretation is as
straight-forward as that. Additionally DPH shows a straight-
forward partition between two phases of 1:1 so we have a
uniform transition. Also the transitions are broad and
close together and so will limit the resolution.

Dr. Dale : They looked rather far apart to me.

Dr. Thulborn : The transitions as seen in the binary
mixture are rather broad as compared to the pure lipid
which is why one can't see them.

Dr. Lee : Are the fluorescence decays from the anthroyl
groups in the bilayer exponential ?

Dr. Thulborn : The data is not yet good enough to allow
analysis for more than one exponential which is why we
use several parameters to analyse a particular property.

Prof. Porter : What are the decay times ?

Dr. Thulborn : They range from 8 to just under 13 ns.

Dr. Tegmo-Larsson : Did you study the fluorescence quenching
effects at high temperatures above $20^{\circ}C$ and if so what are
the quenching effects ?

Dr. Thulborn : This is relevant to the Cu^{2+} work. When
we use temperature and look at the 1% quenching in 2AP we
see only a slight change in the quenching through the
transition of dimyrystoryl phosphatoyl choline vesicles.
When we look at 12AS we see an increase in quenching
around the transition temperature and we believe this is
due to increased permeability of the bilayer to Cu^{2+} as one
goes through the transition temperature which is reminiscent
of permeability profiles as a function of temperature. The
other compounds fall in-between these two curves.

Prof. Chapman : The particular feature of membrane lipids is that the chains like to remain parallel. Adding a probe can make the chain or part of the chains non-parallel and perturb the local environment. An important question is does the probe only give information on these local perturbed environments ? My second point is that the orientation of these probes is not easily predicted *a priori* within the hydrocarbon layer. Furthermore it is possible that the fluorescent probe molecules can aggregate below the lipid transition temperature and disaggregate at higher temperatures. Are the changes you see due to aggregation ? The differing effects of different probes could be due to their different aggregation-disaggregation properties.

Dr. Thulborn : A number of people have worried about this and have used NMR and X-ray scattering to look for aggregation and this certainly occurs in probe concentrations as high as 1:1 or 1:5 with lipid. We looked carefully for perturbation effects and for a system that would show membrane perturbations. One method is to use a BLM on a septum and to measure conductance and capacitance. It is observed that as little as 0.1% of probe can perturb the membrane. So "yes" to your question, if probes change the local region of the bilayer. However, in our experiments, comparison with other techniques in measuring several properties of the membrane gives us confidence in our data.

Prof. Chapman : Some of the larger fluorescent probes must surely perturb the local lipid environment, the more subtle the effect being studied the more there is the possibility of obscuring what you are looking for. This perturbation effect shows up in comparisons made with spin label molecules where large differences can be seen when compared with deuterium NMR studies.

Dr. Thulborn : I believe our technique can be used as a first approximation, one cannot do deuterium NMR on whole cells so probes are a useful attempt in these complex systems to find parameters unobtainable by other techniques.

STEADY STATE DIPHENYL HEXATRIENE
FLUORESCENCE POLARISATION
IN THE STUDY OF CELLS AND CELL MEMBRANES

SHEENA M. JOHNSON [*]

*Medical Research Council Clinical Research Centre,
Watford Road, Harrow, Middlesex HA1 3UJ*

The steady-state fluorescence polarisation characteristics of
1,6-diphenyl-1,3,5,-hexatriene have been used in a convenient
and very sensitive method of comparing the properties of
living cell membranes, isolated cell membranes and model
phospholipid liposomes. Diphenylhexatriene is a rod-shaped
probe which is only soluble and fluorescent in hydrophobic
environments and has been shown to partition into both the
gel and liquid crystalline phases of phospholipid bilayers
in equal proportions (Lentz *et al.* 1976). Phospholipid bilayers and cell membranes are strongly asymmetric, and
Kinosita *et al.*(1977) have shown that the steady-state anisotropy of diphenylhexatriene is governed both by the actual
dynamic viscosity of the environment of the probe in the
membrane and by the orientation constant imposed by the membrane. Hildenbrand and Nicolau (1979) calculated that the
steady-state fluorescence anisotropy r_S is related to r_∞, the
time-resolved fluorescence anisotropy at very long times
after excitation, by the equation:

$$r_S = \frac{r_0 - r_\infty}{1 + \tau/\phi} + r_\infty \tag{1}$$

* Present address : Institut fur Strahlenchemie im Max-Planck-
Institut fur Kohlenforschung, Stiftstrasse 34-36, D-4330
Mulheim a.d. Ruhr, West Germany.

where r_0 is the limiting anisotropy when the probe is immob-
ilised, and τ and ϕ are the probe fluorescence lifetime and
correlation time respectively.

Jahnig (1979) has further analysed the motion of the probe,
and has shown that diphenylhexatriene can be used to detect
lipid chain order, S_ν where ν indicates the mean position of
the probe along the fatty acid chain ;

$$r_\infty = 2/5 \ S_\nu^2 \qquad\qquad (2)$$

Rearranging equation (1)

$$\phi = \{ \frac{r_s - r_\infty}{r_0 - r_s} \} \tau$$

Unfortunately it is not possible to obtain r_∞ from steady
state measurements. If the medium is isotropic, $r_\infty = 0$ and
equation (1) becomes identical to the Perrin equation ;

$$\phi = \frac{r_s}{r_0 - r_s} \ \tau \qquad\qquad (3)$$

For lipid bilayers, values of ϕ calculated from the Perrin
equation will be too high, the error increasing with
increasing order in the bilayer. We use the letter ρ to
distinquish the correlation coefficient calculated from the
Perrin equation from the true correlation coefficient ϕ .
Pending the adaption of our machine to measure time-resolved
fluorescence anisotropy, we have used the Perrin equation to
obtain comparative values of ρ .

Polarisation values for diphenylhexatriene, P, were obtained
directly as the ratio

$$P = \frac{I_\| - I_\perp}{I_\| + I_\perp}$$

using an Elscint microviscosimeter. $I_\|$ was the intensity
of the polarised light emitted by the fluorophor parallel
and $I_\perp$ the intensity of the emitted light polarised perp-
pendicular to the excitation beam. A correction was applied
to compensate for the depolarisation due to light scattering.
P was plotted as a linear function of absorbance of the sample
at 450 nm and the true value of P obtained by extrapolating
to zero absorbance (see Johnson and Nicolau 1977 and fig.1).
Because the ratio is measured directly, the P values obtained

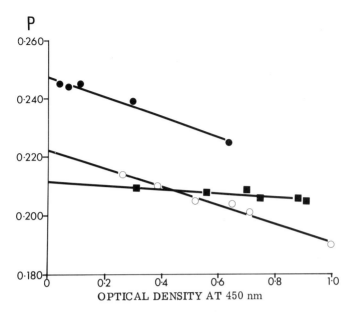

Fig. 1. Values of the degree of fluorescence polarisation, P, versus absorbance at 450 nm of diphenylhexatriene embedded in ■ live lymphocytes, ○ lymphocyte unfractionated homogenate and ● lymphocyte plasma membrane vesicles. All measurements are at 37°C, λ_{ex} = 365 nm.

are independent of fluctuations in the intensity of the light source, and each point was usually accurate to about ± 0.002.

Fluorescence lifetimes were measured with an Applied Photophysics time correlated single photon counting nanosecond spectrometer, and a computer was used to convolute the best fit single exponential decay curve to the experimental data, starting from the lamp profile. The variance between the convoluted and experimentally measured decay curve was very little greater than that measured for the known single exponential fluorescence decay of 9-cyanoanthracene in ethanol (Dale *et al.* 1977). We have found that if the lipids were oxidised the fluorescence lifetime of diphenylhexatriene was shorter, and P greater, but there was no change in ρ , the correlation time calculated from the Perrin equation.

We initially became interested in the use of the fluorescence characteristics of diphenylhexatriene in the study of membranes when Inbar and Shinitzky and coworkers reported that human and mouse leukaemic and lymphoma cells had a lower diphenylhexatriene polarisation value and were more fluid than normal lymphocytes. They attributed this to a lower

cholesterol/phospholipid ratio in the plasma membranes of the
pathological lymphocytes. We have not been able to confirm
the observation of Inbar and Shinitzky with human cells, see
table 1.

TABLE 1
Diphenylhexatriene Polarisation Values for
Human Lymphocytes at 37°C

Cell	P	
Normal cells :		
Circulating lymphocytes + platelets + monocytes	0.251 ± 0.003	(3)
Circulating lymphocytes + 20% monocytes	0.234 ± 0.002	(6)
Circulating T lymphocytes	0.224 ± 0.010	(4)
Tonsil lymphocytes	0.214 ± 0.010	(19)
Tonsil T lymphocytes	0.218 ± 0.006	(5)
Pathological cells :		
Chronic lymphocytic leukaemia	0.224 ± 0.008	(17)
Acute lymphoblastic leukaemia	0.212 ± 0.009	(3)
Lymphosarcoma	0.227 ± 0.007	(2)

Results are shown as the mean $\pm$ standard deviation, with the
number of samples in brackets.

The problem appeared to be one of cell purity. In general,
patients with untreated lymphocytic leukaemias have a very
large number of lymphocytes in their blood so that it is
possible to obtain a virtually pure lymphocyte preparation.
Healthy people have a much lower proportion of lymphocytes,
and these have to be carefully separated from other leuco-
cytes and platelets before a satisfactory preparation can be
obtained. Mononuclear cells containing platelets from normal

blood had a diphenylhexatriene polarisation value of 0.251
± 0.003 in good agreement with the value for 'normal lymphocytes' quoted by Inbar and Shinitzky. However this figure
became less as the lymphocytes were successively freed from
contaminating cells. If the blood was defribinated to remove
platelets the mononuclear cells (lymphocytes and monocytes)
from five individual donors had a polarisation value of
0.237 ± 0.006. Purified T cells had a still lower polarisation of 0.224, identical to that of the chronic lymphocytic leukaemic cells and not significantly different from
the value of 0.218 of the normal tonsil T lymphocytes
(Johnson and Kramers,1978).
     We were able to confirm the observations of Inbar and
Shinitzky on some mouse leukaemic lines. In this case large
numbers of both normal and pathological lymphocytes can be
obtained from the lymphoid tissues of these animals, and the
polarisation values for diphenylhexatriene were lower in the
pathological cells.
     We did not confirm that this difference was due to less
cholesterol in the plasma membranes of the leukaemic cells.
     Preliminary experiments showed that diphenylhexatriene was
not confined to the outside(plasma) membrane of the cells.
We compared the fluorescence intensity of diphenylhexatriene
in subcellular fractions to conventional enzyme markers, and
showed that the distribution of the fluorescence was similar
to that of the phospholipids, but not to the two markers of
the plasma membrane, cholesterol and 5'-nucleotidase (see
table 2, Johnson and Nicolau, 1977).

TABLE 2

*Percentage distribution of the fluorescence of diphenylhexa-
triene compared to the distribution of phospholipid, 5'-
nucleotidase and cholesterol in human tonsil lymphocytes
subject to differential centrifugation.*

Sample (pellet)	Membrane DPH fluorescence	Phospho-lipid (mole %)	5'-nucleo-tidase activity	Cholesterol (mole %)
300 g	19.0	19.0	11.3	12.6
4.000 g	24.5	26.1	22.5	19.1
20.000 g	24.1	22.4	34.3	35.3
Supernatant	32.3	32.5	31.9	33.0
Recovery	92.4	112.7	102.7	106.4

TABLE 3

Diphenylhexatriene correlation times at 37° for isolated lymphocyte plasma membrane vesicles and liposomes prepared for, then compared to, the cholesterol/phospholipid ratios

Lymphocyte	Sample	P	τ (ns)	ρ	Plasma membrane cholesterol/phospholipid molar ratio		
					Biochemical (experimental)	DPH fluorescence	Pure membrane (calc.)
Human tonsil							
	PM preparation	0.244	8.6	8.2	–	–	–
	PM liposome	0.214	7.75	5.7	0.42 ± 0.03 (6)	0.42	0.49
Human CLL							
	PM preparation	0.248	8.3	8.2	–	–	–
	PM liposome	0.205	8.2	5.6	0.37 ± 0.04 (5)	0.375	0.50
Mouse spleen							
	PM preparation	0.245	8.0	7.7	–	–	–
	PM liposome	0.226	8.3	6.8	0.41 ± 0.02 (2)	0.425	–
Mouse Gardner lymphoma							
	PM prepation	0.193	8.0	4.9	–	–	–
	PM liposome	0.175	7.3	3.8	0.34 ± 0.07 (7)	0.38	0.45

We therefore isolated purified plasma membranes from normal
and pathological lymphocytes of mouse and man, and found their
cholesterol/phospholipid ratios both by chemical and biochemical methods, and by measuring the diphenylhexatriene correlation coefficients. The contamination of the plasma membrane
preparation by other cell fragments was carefully checked by
assaying marker enzymes, and by measuring the proportions of
the membranes in the original cell. This allowed us to
calculate the cholesterol/phospholipid ratio for the pure
membrane. The results are shown in table 3 (Johnson and
Robinson,1979).
     For the mouse lymphoma, it will be noticed that the diphenyl-
hexatriene correlation time is lower in both the plasma
membrane and plasma membrane liposomes compared to normal
mouse cells. The cholesterol/phospholipid ratio is almost the
same. This is due to an increase in the triglycerides and
alkyl diacylglycerides in the lymphoma cells, compared to the
normal mouse or the human lymphocytes, see table 4.

TABLE 4

*Total Glycerides present in Lymphocyte
Plasma Membrane*

Lymphocyte	Total Glycerides *
Human tonsil	0.079
Human chronic lymphocytic leukaemia	0.051
Mouse spleen	0.030
Mouse Gardener Lymphoma	0.179

* The results are glyceride/phospholipid molar ratios.

     The effect of the triglyceride, triolein, on the diphenyl-
hexatriene correlation time in model liposomes is shown in
figure 2. The liposomes contained approximately the same
phospholipids as were found in the lymphocyte plasma membrane.
Triolein has little effect on the correlation time of
diphenylhexatriene in the pure phospholipids, but apparently
reverses the effect of cholesterol.
     We studied the effects of triglycerides further on liposome
systems. The results indicated that most of the triglycerides

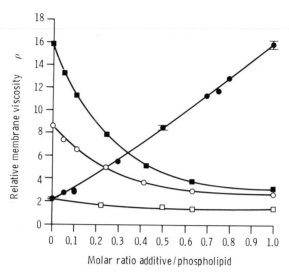

*Fig. 2. The correlation coefficient for model phospholipid/
cholesterol membranes.* ● *phospholipid + increasing mol. ratio
of cholesterol. The bars represent the S.D. on five different
preparations ;* □ *phospholipid + increasing mol. ratio of
triolein;* ○ *phospholipid:cholesterol (2:1) + increasing
mol. ratio of triolein;* ■ *phospholipid:cholesterol (1:1)
+ increasing mol. ratio of triolein.*

were located in a separate region in the liposome, probably
in the centre of the bilayer as indicated in figure 3. We
found that the presence of triolein lowered the solid/liquid
crystalline phase transition of dipalmitoylphosphatidylcholine by only 0.9°C and that it did not affect the ability

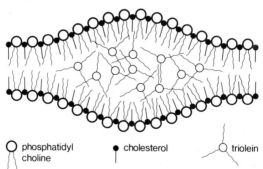

*Fig. 3. Suggested arrangement for triolein in a phosphat-
idyl choline bilayer, with or without cholesterol.*

of cholesterol to abolish the phase transition, see figure 4.
On the other hand, if we use trielaidin, the trans isomer of
triolein, we were able to obtain two phase changes, one from
the dipalmitoylphosphatidylcholine which showed no evidence
of supercooling and was abolished by cholesterol, and one
which did show supercooling and was not affected by cholesterol (see figure 5).

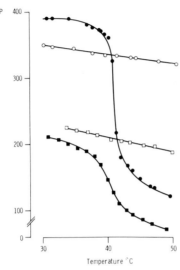

*Fig. 4. The polarisation, P. of diphenylhexatriene in lipo-
somes containing equimolar ratios of dipalmitoylphosphatidylcholine and triolein with and without cholesterol as a function of temperature. Cooling and heating curves were
identical. No Transition was observed.*

● *Dipalmitoylphosphatidyl choline*

○ *1:1 Cholesterol:dipalmitoylphosphatidyl choline*

□ *1:1:1 Cholesterol:triolein:dipalmitoylphosphatidyl-
choline*

■ *1:1 Triolein:dipalmitoylphosphatidyl-choline*

Negatively stained electron micrographs of the liposomes
showed an irregular thickening in the bilayer when triglycerides were present (see figure 6).
    Consistent with these findings, cells which have a lot of
triglycerides normally store it as droplets in the cytoplasm.
Pessin *et al.* (1978) and van Hoeven *et al.* (1979) have
reported low cell diphenylhexatriene polarisation results
correlating with the presence of liquid droplets in the cell.
We did not see any liquid droplets in the initial electron

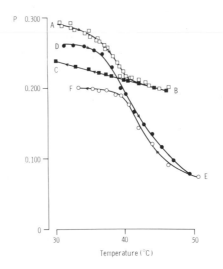

Fig. 5. *The polarisation, P. of diphenylhexatriene in lipo-*
somes containing equimolar ratios of dipalmitoylphosphatidyl-
choline and trielaidin, with and without cholesterol as a
function of temperature. Curve ABC 1:1:1 Trielaidin:choles-
terol:dipalmitoylphosphatidyl-choline.

   □ *Curve AB, heating after storing at 4° for 1 hour*

   ■ *Curve BC, cooling*
Curve DEF 1:1 Trielaidin:dipalmitoylphosphatidyl-choline.

   ● *Curve DE, heating after storing at 4° for 1 hour*

   ○ *Curve EF, cooling*

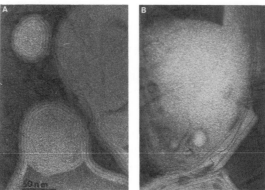

Fig. 6. *Electron micrograph of phosphatidyl-choline vesicles*
(A). Note the irregular bilayers when triolein is present
as a 1:1 molar ratio (B).

micrographs of the lymphocytes listed in table 4, so perhaps this amount of triglyceride can be included in the plasma membrane. The Gardner lymphoma however becomes increasingly glyceride rich on successive passages through mice and it is possible that some of the later cells included lipid droplets.

In conjunction with Nicolau, Friis and coworkers (1977) we investigated the behaviour of diphenylhexatriene embedded in the membranes of avian fibroblasts. Normal chicken or quail fibroplasts showed an increase in the diphenylhexatriene polarisation and correlation time when the cells formed a confluent layer and stopped growing. Oncogenically transformed cells which can form tumours lose their growth control mechanism, and these showed no increase in diphenylhexatriene correlation time when the cells became confluent. We were able to show that the change in correlation time between non-confluent and confluent cells was not due to a change in the proportion of internal cell membranes and could be measured also in isolated plasma membranes. Two types of transformation were studied. In one case the cells were transformed with the Rous sarcoma virus, and in the second case we use a methyl cholanthrene derived tumour cell line, one strain of which had regained its growth control mechanism and so lost its ability to form tumours. The results are summarised in table 5.

TABLE 5

Rotational correlation time in nanoseconds of diphenylhexatriene embedded in whole cells and isolated plasma membrane vesicles at 37°C

Fibroblast		Whole cells	Plasma membranes
Normal quail	Subconfluent	3.3 ± 0.6	5.1 ± 0.6
	Confluent	6.9	8.3
Virus transformed	Subconfluent	3.5	6.3
	Confluent	3.5	6.2
Non-tumourogenic	Subconfluent	5.9	6.0
Methyl cholan-threne cells	Confluent	6.6	7.6
Tumourgenic	Subconfluent	4.8	5.8
Methyl cholan-anthrene cells	Confluent	4.8	5.7

Experimentally we found that the lifetime of diphenylhexatriene embedded in mixtures of phospholipid, cholesterol or
triolein was approximately a linear function of the polarisation (see figure 7).

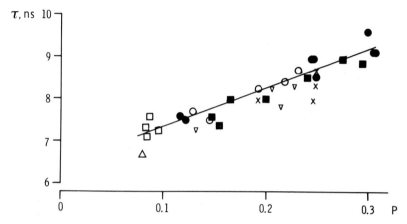

Fig. 7.   The empirical relation between lifetime and polar-
isation for diphenylhexatriene embedded in liposomes or
cell membranes.   The regression line was drawn through all
the model lipid liposomes and has the form :

$$\tau \quad = \quad 9.28 \ P \quad + \quad 6.42$$

The phospholipid, PL, composition was 46% phosphatidyl
choline, 29% phosphatidyl ethanolamine, 12% phosphatidyl
serine and 13% sphingomyelin. □ *PL + triolein ;*
● *PL + cholesterol;* ■ *PL:cholesterol 1:1 + triolein;*
○ *PL : cholesterol 2:1 + triolein ;* △ *triolein alone ;*
✕ *lymphocyte plasma membranes ;* ▽ *lipid extract from*
lymphocyte plasma membranes.

If diphenylhexatriene is present in a hydrophobic region
in a protein, the lifetime is short compared to the high P
value and the decay of fluorescence does not fit a single
exponential. For example, luciferase at 33⁰ had a P value
of ~ 0.3 and a best fit 'single exponential' lifetime of
5.0 ns. It will be seen that in real plasma membranes
diphenylhexatriene behaves as if it is located in the lipid
region, although this appears to be stiffened by the membrane
proteins.

REFERENCES

Dale R.E., Chen L.A. and Brand L. (1977) *J. Biol. chem.*,
__252__, 7500-7510
Hildenbrand K and Nicolau, C. (1979) *Biochim. Biophys. Acta*
553, 365-377.
Inbar M., Shinitzky, M. and Sachs, L. (1974) *Febs Letters*
__38__, 268-270
Inbar, M. and Shinitzky, M. (1974) *Proc. Nat. Acad. Sci. USA*
71, 4229-4231
Jahnig, F. (1979) *Proc. Natl. Acad. Sci. USA* __76__, 6361-6365
Johnson, S.M. and Nicolau, C. (1977) *Biochem. Biophys. Res.*
Commun. 76, 869-874
Johnson, S.M.and Kramers, M. (1978)   *Biochem. Biophys. Res.*
Commun. __80__, 451-457
Johnson, S.M. and Robinson, R. (1979) *Biochim. Biophys. Acta*
__558__, 282-295
Kinoshita, K., Kawato, S. and Ikegami, A. (1977) *Biophys. J.*
20, 289-305
Lentz, B.R., Barenholz, Y. and Thompson, T.E. (1976)
Biochemistry 15, 4529-4537
Nicolau, C., Hildenbrand, K., Reimann, A., Johnson, S.M.,
Vaheri, A., and Freiis, R. (1978) *Exp. Cell Res.*113, 63-73
Pessin, J.E., Salter, D.W. and Glaser, M. (1978)*Biochemistry*
17, 1997-2004
Shinitzky, M. and Inbar, M. (1976) *Biochim. Biophys. Acta*
433, 133-149
Van Hoeven, R.P., van Blitterswijk, W.J. and Emmelot, P.
(1979) *Biochim. Biophys. Acta* 551, 44-54

DISCUSSION ON DR. JOHNSON'S PAPER

Dr. Thulborn : You mentioned that pre-transition in the phospholipid bilayer was a head group effect. In fact it is now known to be a hydrocarbon effect. I am not sure why you do not see this transition in your data.

Dr. Johnson : We have looked carefully for this transition but have not observed it in all cases.

Prof. Chapman : The pre-transition depends on the rate of cooling and also on the absence of impurities such as too many probe molecules themselves.

Dr. Johnson : We carefully checked the purity of the sample and the melting points were as they should be.

Dr. Thulborn : Did you heat or cool the sample to approach the transition ? One should see the pre-transition first on heating taking 20 or so minutes to cover the temperature range.

Dr. Johnson : We have looked for the pre-transition both on cooling and heating taking 1 hour to complete the transition.

Prof. Chapman : Can you tell me where the probe is situated and how it is orientated in the membrane ?

Dr. Johnson : I understand that these DPH probes go into the hydrocarbon region of the lipid with the long axis of the triene parallel or nearly so with the lipid hydrocarbon chains.

Dr. Dale : Some earlier work shows that the probes are perpendicular to the plane of the membrane (Badley *et al. Biochemistry* (1973), 12, 268).

Prof. Porter : What model do you have for DPH in the membrane ?

Prof. Chapman : Some workers suggest that DPH may be parallel to the surface and in the middle of the bilayer.

Dr. Johnson : One would not expect DPH to be sensitive to the presence of cholesterol which is near the head groups.

Prof. Chapman : Do the probes aggregate below the phase transition ?

Dr. Dale : This is difficult to tell as they interact only weakly.

Prof. Porter : Then why would they aggregate ?

Dr. Dale : "Aggregate' was used in the sense of freezing-out rather than complex formation.

Prof. Porter : Could one look at the dichroism of DPH ?

Dr. Dale : There are already some reports in the literature (Cehelnik *et al.*, *Proc. Roy. Soc.*, (1973), A335, 387)

Prof. Chapman : To return to probes in biomembranes - spin probes sometimes enter crevices of proteins and one sees protein events not necessarily those associated with lipid phase changes. Could the same behaviour be occurring with DPH i.e. with these cells is one observing protein not lipid effects ?

Dr. Johnson : In practice only a little DPH enters the protein but sometimes a probe may enter any hydrophobic part of the protein.

Dr. Tegmo-Larsson : How did you prepare the liposomes since this might show a variation in the location of DPH. Were the vesicles multiple or single ?

Dr. Johnson : The lipid, DPH and chloroform were mixed and the solvent removed in a rotating evaporator under nitrogen. The buffered water was added and the mixture shaken vigorously. Sonification was not used so all the liposomes were multilayered.

Dr. Tegmo-Larsson : Could the vesicle's properties be different because of the strain introduced by their small size ?

Prof. Chapman : NMR, and other techniques, show very little difference between vesicles and liposomes.

Dr. Johnson : Doesn't phosphorus NMR show that the inside and outside of the vesicle membranes are different ?

Prof. Chapman : There is no real evidence of strain of the bilayer in sonicated vesicles or that the inside and outside are vastly different from the point of view of deuterium NMR studies to study order parameters or relaxation times.

Dr. Lee : There is a big change in the phase transition on sonication as it is widened.

Dr. Johnson : What is the pre-transition ?

Prof. Chapman : This is a hydrocarbon chain event but may also involve the polar group of lecithin.

Dr. Quinn : In NMR studies of pre-transitions, the chains change by 23^0 from being almost perpendicular to the surface possibly due to a change in the water molecule's penetration.

Prof. Chapman : The pre-transition is certainly a real effect since it is connected to ripples observed in freeze fracture preparations observed by electron microscopy and to a particular phase of the lipid.

APPLICATION OF FORSTER LONG-RANGE EXCITATION ENERGY TRANSFER TO THE DETERMINATION OF DISTRIBUTIONS OF FLUORESCENTLY-LABELLED CONCANAVALIN A-RECEPTOR COMPLEXES AT THE SURFACES OF YEAST AND OF NORMAL AND MALIGNANT FIBROBLASTS[*]

ROBERT E. DALE,[+] JOEL NOVROS,[+] STEPHEN ROTH[++]

MICHAEL EDIDIN[++] and LUDWIG BRAND[++]

Department of Biology and McCollum-Pratt Institute
The Johns Hopkins University, Baltimore
Maryland, 21218, U.S.A.

ABSTRACT

The mitogenic lectin concanavalin-A binds to cell surface glycoproteins and can initiate events leading to cell division. The distribution and/or redistribution of the initial complexes over the plasma membrane may play a decisive role in this process and may differ between normal and oncogenically transformed (malignant) cells. A potentially powerful method of examining such distributions *in vivo* is provided by Forster long-range (resonance) excitation energy-transfer between a suitable pair of fluorescent donor and acceptor labels attached to separate concanavalin-A molecules. The energy transfer

---

[*] This work was supported by National Institutes of Health Grant GM 11632 and by National Science Foundation Grant GB 37555. An extended abstract has already appeared in the literature (Nicolson, 1972).
[+] Present address : Paterson Laboratories, Christie Hospital and Holt Radium Institute, Manchester M20 9BX.
[+] Present address : National Health Laboratories, 1007 Electric Avenue, Vienna, Virginia 22180, U.S.A.
[++] Supported by National Institutes of Health Career Development Awards GM 10245 (L.B.) and HD 00148 (S.R.)

efficiencies between donor fluorescein- and acceptor tetra-
methylrhodamine-derivatised concanavalin A's adsorbed onto
cultured BALB/c 3T3 murine fibroblasts and onto the spontaneously and virally transformed derivatives, 3T12 and SV3T3
respectively, as also onto the cell wall of yeast as a model
system, were determined. The results, whose quantitative
interpretation required detailed consideration of a number
of non-trivial data correction procedures, are discussed in
relation to existing electron microscopic data in the literature.

INTRODUCTION

Why Study Distributions on Cell Surfaces ?

Many cellular processes may be triggered or controlled *via*
events at the external surface of the cell involving intrinsic
plasma membrane proteins, often glycoproteins. Density-
dependent growth control, i.e. inhibition of further cell
growth and division on cell-cell contact, a property lost on
oncogenic transformation, is probably an extended example of
this. Conversely, triggering of resting cells into division
by externally bound agents such as the mitogenic lectins
is a widely studied phenomenon of this type. Again, hormones
and antibodies that may react very specifically with cell
surface components (receptors) can also mediate changes in
cell behaviour and metabolism without even entering the cell.
  The initial distibution and changes in distribution by
lateral diffusion over the cell surface following binding
of these materials may be important for co-operative mediation
of these effects. Similarly the mutual topography of two or
more different effector-receptor complexes in concerted or
opposing roles may help provide an even higher degree of
subtlety to the sensitive regulatory apparatus. Knowledge
of such topographical characteristics and their dynamics
may therefore aid in understanding the detailed biochemical
and biophysical mechanics of these effects and pinpoint
where they may be aberrant in pathological conditions such
as malignancy.

Why Study Distributions on Cell Surfaces by Energy Transfer ?

The advantages that may accrue from the use of Forster long-
range (resonance) excitation energy-transfer, FRET (Forster,
1965), to study the topography of cell surface components
can be summarised under five headings :

(a) in common with all fluorescence methods, it is very

sensitive in that potentially very small quantities of material
may be examined, e.g. single cells or even organelles containing picomoles or even smaller quantities of fluorescent
material;

(b) due to the basis dependence of the transfer rate on the
inverse sixth power of the separation between donor and
acceptor moities, FRET is potentially very sensitive to
proximities in the nanometer range - just the order of
magnitude of separations that may be expected in the system
being considered here;

(c) the method is potentially highly selective in that
particular components of interest may be open to specific
fluorophoric labelling;

(d) the dynamics of topographical redistribution can potentially be followed on any time scale down to at least the
nanosecond range, possibly even further;

(e) finally, and perhaps most importantly,compared with other
techniques such as section or scanning electron microscopy
of whole or fixed or of freeze-fractured preparations, the
method is potentially a very gentle one, capable of *'in vivo'*
application.

ATTAINMENTS AND PROSPECTS ; AIMS OF THE PRESENT STUDY

That fufillment of the promise of the FRET method in this
area is a realistic goal can be seen in recent reports in
the literature of a number of studies on model membrane
systems (Fung and Stryer, 1978; Estep and Thompson, 1979;
Koppel *et al.*, 1979), on erythrocyte 'ghosts' (Shaklai*et al.*,
1976a,b), and also in studies on whole-cell plasma membranes
(Fernandez and Berlin, 1976; Chan *et al.*, 1979) parallel to
those which will be described briefly below. As indicated
above, the general aim of such studies on the living cell
is to sensitively and accurately quantitate the absolute
and/or relative distributions of suitably fluorescently-
labelled biologically interesting components of its plasma
membrane. The specific aims of the work reported here were :

(1) to establish the technique in a relatively simple case,
that of the fluorescently-labelled mitogenic lectin concanavalin-A (conA) adsorbed into the thick proteoglycan cell
wall of yeast in large amounts, thus providing a model
system in which a random distribution and high energy transfer
efficiency may be reasonably assumed and checked and

(2) to study quantitatively the distribution of conA-receptor complexes on the surfaces of normal and transformed fibro-blasts, there being conflicting evidence in the literature from electron microscopic studies as to the randomicity or otherwise of their distribution under a variety of conditions, and as to whether or not the distributions may differ between normal cells and their malignant derivatives (Nicolson, 1971, 1972; de Petris *et al.*, 1973; Collard and Temmink, 1975).

In the following, the theoretical basis for such measure-ments (Forster, 1949) will be summarised. A limit and an extension of the theory appropriate to the cases under consideration will also be presented, along with a brief outline of the experimental protocol and the results (a full report on the experimental techniques and results will be given elsewhere, Novros *et al.*, in preparation).

THEORETICAL FOUNDATIONS

The basis of the FRET method is that the rate of transfer between an isolated donor (D)-acceptor (A) pair may be written (Forster, 1965) as :

$$k_T = (1/\tau_0)(R_0/R)^6 \qquad (1)$$

where τ_0 is the inverse of the first order decay rate const-ant of the excited donor in the absence of acceptor and R the separation between D and A. R_0 is the characteristic separation defined by the spectroscopic properties of the system:

$$R_0^6 = C \kappa^2 \phi_0 n^{-4} J \qquad (2)$$

where C is a combination of physical and mathematical constants, κ^2 an orientation factor, ϕ_0 the quantum yield of the donor in absence of the acceptor, n the effective refractive index of the intervening medium and J an overlap integral expressing the degree of 'resonance' between D emission and A absorption spectra :

$$J = \int_0^\infty \varepsilon(\lambda)F(\lambda)\lambda^4 \, d\lambda / \int_0^\infty F(\lambda) \, d\lambda \qquad (3)$$

where $\varepsilon(\lambda)$ is the molar extinction coefficient of the acceptor and $F(\lambda)$ the relative donor emission intensity (photons per unit wavelength interval) at wavelength λ. With $\varepsilon(\lambda)$ in $cm^2/mmole$ and λ in cm, the characteristic separation is given by :

$$R_o = 9.786 \times 10^{-5} (\kappa^2 \phi_o n^{-4} J)^{1/6} \text{ cm} \quad (4)$$

For the isolated D-A pair, under the condition that there is no appreciable back-transfer (which will occur if the possible A emission spectrum overlaps the D absorption spectrum to any extent - a situation which can usually be avoided, as in the present case of fluorescein donor and tetramethylrhodamine acceptor), the decay course of D in the presence of A remains exponential :

$$\rho(t) = \exp\{-(\tau_o^{-1} + k_T) t\} \quad (5)$$

The energy transfer efficiency is given by the ratio of the rate of transfer to that of all deactivation processes including transfer :

$$\phi_T = k_T / (\phi_o^{-1} + k_T) \quad (6)$$

or equivalently, by the ratio of the quantum yields of D emission in presence or absence of A:

$$(\phi/\phi_o) = \int_0^{\infty} \rho(t) dt \bigg/ \int_0^{\infty} \rho_o(t) dt$$

$$= 1 - \phi_T \quad (7)$$

where $\rho_o(t) = \exp\{-t/\tau_o\}$.

APPLICATION TO TWO- AND THREE-DIMENSIONAL DISTRIBUTIONS

Of concern in the present application is the transfer occurring between many donors and acceptors which are supposed to be randomly (isotropically) distributed throughout the available two- or three-dimensional space in which they find themselves. Provided that, in addition to the other str ctures indicated above, self-transfer between donors, which arises because of some overlap between the absorption and emission spectra of D, can be neglected, the transfer efficiency observed will be an average over all possible spatial arrangements of an acceptor population about a donor. This problem has long since been solved both for three-dimensional (Forster, 1949; Birks and Georghiou, 1968) and two-dimensional (Tweet *et al*., 1964; Hauser *et al*., 1976; Wolber and Hudson, 1979) distributions in the limit of negligible translational by rapid (dynamic) reorientational

diffusion, as well as for the negligible (static) reorientational averaging regime in the 3-D case (Galanin, 1955;
Maksimov and Roznan, 1962; Steinberg, 1968). In these
treatments, the time-dependence of the donor decay is
determined analytically, and numerical evaluation of the
integral over time used to obtain the decrease in donor
quantum yield and thereby the transfer efficiency, assuming
that no additional quenching of D emission over and above
that due to FRET is induced by presence of acceptors (Schiller,
1975).

Thus the donor decay for the 2-D acceptor distribution is
given by :

$$\rho_2(t) \quad = \quad \exp\left|-\{(t/\tau_0) + p(t/\tau_0)^{1/3}\}\right| \quad (8)$$

and analogously for the 3-D case by :

$$\rho_3(t) \quad = \quad \exp\left|-\{(t/\tau_0) + q(t/\tau_0)^{1/2}\}\right| \quad (9)$$

On substituting $x = t/\tau_0$ and applying eqn.7, the corresponding
transfer efficiencies may be obtained as :

$$\phi_{T2} \quad = \quad 1 - \int_0^\infty \exp\{-(x + px^{1/3})\}dx \quad (10)$$

and :

$$\phi_{T3} \quad = \quad 1 - \int_0^\infty \exp\{-(x + qx^{1/2})\}dx \quad (11)$$

The constants p and q appearing in these equations are
related to the concentrations (volume and surface respectively) of acceptor :

$$p \quad = \quad \Gamma(2/3)(R_0/\bar{R}_2)^2$$

$$\simeq \quad 1.354 \ (R_0/\bar{R}_2)^2 \quad (12)$$

and :

$$q \quad = \quad \Gamma(1/2)(R_0/\bar{R}_3)^3$$

$$= \quad \pi^{1/2}(R_0/\bar{R}_3)^3 \quad (13)$$

where the gamma function is defined by :

$$\Gamma(n) \quad = \quad \int_0^\infty e^{-z} \, z^{n-1} dz \qquad (14)$$

and the values indicated may be obtained from standard tables using the recursive relationship :

$$\Gamma(n + 1) \quad = \quad n\Gamma(n) \qquad (15)$$

The mean values $\overline{R}_2$ and $\overline{R}_3$ correspond to the radii of the disc and sphere respectively containing, on average, a single acceptor molecule so that :

$$c_2 \quad = \quad 1/(\pi N' \overline{R}_2{}^2) \qquad (16)$$

and :

$$c_3 \quad = \quad 3/(4\pi N' \overline{R}_3{}^3) \qquad (17)$$

where the 2-D and 3-D concentrations c_2 and c_3 are defined in mmole per cm^2 and cm^3 respectively and N' is Avogadro's number per mmole. 2-d and 3-D 'critical' concentrations may be defined in the same way using R_0 in place of $\overline{R}_2$ and $\overline{R}_3$ respectively so that the ratios in eqns. 12 and 13 become (c_2/c_{02}) and (c_3/c_{03}) respectively.

The energy transfer efficiencies for the two cases can be determined analytically to any desired degree of accuracy by power series expansion of the second of the two exponential functions contained in each of the two integrals of eqns. 10 and 11 and application of the gamma function defined in eqns. 14 and 15 (Wolber and Hudson, 1979), leading to :

$$\phi_{T2} \quad = \quad 1 \; - \; \sum_{k=0}^{\infty} \{(-p)^k \Gamma(k/3 + 1)/k!\} \qquad (18)$$

and :

$$\phi_{T3} \quad = \quad 1 \; - \; \sum_{k=0}^{\infty} \{(-q)^k \Gamma(k/2 + 1)/k!\} \qquad (19)$$

The transfer efficiencies evaluated from these equations to a tolerance of $< \pm 0.001$ are displayed in figure 1 as a function of both $(R_0/\overline{R}_2)$ or $(R_0/\overline{R}_3)$ and of the reduced concentration (c_2/c_{02}) or (c_3/c_{03}) respectively.

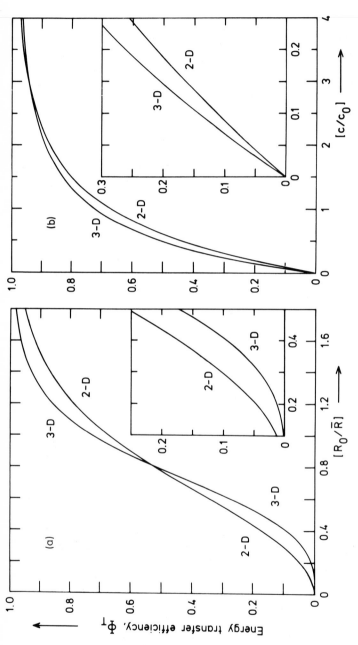

Fig. 1. Forster resonance energy transfer (FRET) efficiencies for stationary statistical (random) acceptor distributions in two and three dimensions. (a) efficiency as a function of the ratio $(R_0/\bar{R}_n)$, where $\bar{R}_n$ is the radius of the disc or sphere containing on average 1 acceptor moiety in $n=2$ and 3 dimensions respectively. (b) efficiency as a function of the reduced concentration $(C_n/C_{0n}) = (R_0/\bar{R}_n)^n$. Insets : expanded scale representation for low transfer efficiencies.

EFFECT OF REORIENTATIONAL AVERAGING REGIME

The values of R_0 in eqns. 12 and 13 will not, in general, be identical. The appropriate average value of the orientation factor κ^2 appearing in $R_0{}^6$ depends on (i) the allowed range of mutual orientations which may be restricted, particularly in the 2-D situation, e.g. Tweet *et al.*, 1964, and (ii) the reorientational averaging regime. The orientation factor may be expressed as :

$$\kappa^2 = (1 + 3 \cos^2\psi) \cos^2\theta \qquad (20)$$

where ψ is the angle between the transition moment associated with D emission and the separation vector, while θ is that between the electric field vector produced by the excited donor at A and the transition moment associated with A absorption (Steinberg, 1968).

Assuming that the orientations of D and A are isotropically distributed in three dimensions (as will obtain in the case presently under consideration for both 3-D and 2-D distributions of the acceptor moities themselves), the appropriate value in the dynamic reorientational averaging regime (rates of reorientation much faster than transfer rates) is given by :

$$\langle\kappa^2\rangle = \frac{\int_0^\infty (1 + 3 \cos^2\psi) \sin\psi d\psi \int_0^\infty \cos^2\theta \sin\theta d\theta}{\int_0^\Pi \sin\psi d\psi \int_0^\Pi \sin\theta d\theta} \qquad (21)$$

where the sine terms represent the weighting element for a given angle. On substituting $x = \cos\psi$ and $y = \cos\psi$, this reduces to :

$$\langle\kappa^2\rangle = \int_0^1 (1 + 3x^2) \, dx \int_0^1 y^2 \, dy = 2/3 \qquad (22)$$

the usually quoted dynamic random (isotropic) value. In this case, the orientation factor has this value for each and every acceptor in the distribution.

In the static random averaging regime, however, there is a spread of κ^2 values as well as of separations, over which the donor decay in the presence of acceptors requires to be averaged. It turns out that the values of p and q should contain the average of an appropriate power of κ (Galanin, 1955; Maksimov and Roznan, 1962; Steinberg, 1968). Inspection of eqns. 12 and 13 defining p and q indicates that the average values appearing in $R_0{}^6$ should be $(\langle\kappa^{2/3}\rangle)^3$ and $(\langle\kappa\rangle)^2$

respectively. As has been shown the latter is an analytical
function :

$$<\kappa> \ = \ \int_0^1 (1 + 3x^2)^{1/2} \ dx \ \int_0^1 y \ dy \ \underset{\sim}{} \ 0.690 \qquad (23)$$

so that κ^2 is replaced in eqn. 2 for R_0^6 by 0.476 $\underset{\sim}{}$ $(0.690)^2$
in the case of a 3-D distribution of acceptors (Galanin,
1955; Maksimov and Roznan, 1962; Steinberg, 1968). Similarly,
for the 2-D acceptor distribution:

$$<\kappa^{2/3}> \ = \ (3/5) \int_0^1 (1 + 3x^2)^{1/3} dx \qquad (24)$$

which affords the value of 0.740 on numerical evaluation
leading to 0.405 as the appropriate value of κ^2 and R_0^6 for
this case.

EFFECT OF A SECOND ACCEPTOR SPECIES

The addition of a second acceptor population also distributed
statistically throughout the available 2-D or 3-D space, a
possibility that is realised in the present described experimental situation, can readily be incorporated into the
expressions given in the previous section by a trivial
extension of the original derivation (Forster,1949). The
contributions of each acceptor population to the donor decay,
and thereby also to the overall transfer efficiency, partition
in a simple way leading to replacement of the values p and
q in eqns. 12 and 13 respectively by :

$$p \ \underset{\sim}{} \ 1.354 \ \{(R_{01}/\overline{R}_{21})^2 \ + \ (R_{02}/\overline{R}_{22})^2\} \qquad (25)$$

and

$$q \ = \ \pi^{1/2} \ \{(R_{01}/\overline{R}_{31})^3 \ + \ (R_{02}/\overline{R}_{32})^3\} \qquad (26)$$

where $\overline{R}_{ij}$ are related to the appropriate i-dimensional
concentration c_{ij} of the acceptor species j by eqns 16 and
17.

When the mole ratio $m_{i2} = (c_{i2}/c_{i1})$ of the two acceptor
populations is known, p and q may conveniently be redefined
on the basis of the absolute concentration of one of them,
e.g. species 1, by defining effective R_0 values :

$$R_{0,eff}^i \ = \ (R_{01}^i \ + \ m_{i2}R_{02}^i) \qquad (27)$$

for substitution with $\underline{i}$ = 2 and 3 into eqns. 12 and 13
respectively in which R_i refers now to acceptor species 1.
The generalisation of eqn. 27 to n different acceptor species:

$$R^i_{0,eff} = (R^i_{01} + \Sigma^n_{j=2} m_{ij}R^i_{0j}) \qquad (28)$$

with $m_{ij} = (c_{ij}/c_{i1})$ follows immediately. Using the appropriate effective R_0 values in eqns. 12 or 13, the <u>overall</u>
energy transfer efficiency is given by eqn. 10 or 11
respectively, and the relative donor quantum yield then
follows from eqn. 7.
     Of the energy thus transferred, a constant fraction at any
given time (and therefore over all time) is transferred to
each of the acceptor species j. Using the generalised
formulation:

$$\phi_{Tij} = \{R^i_{0j} /(R^i_{01} + \Sigma^n_{j=2} m_{ij}R^i_{0j}) \phi_{Ti} \qquad (29)$$

In general, the overall transfer efficiency ϕ_T will be
measurable as a change in quantum yield of the donor, while
ϕ_{Tj} are obtainable from sensitisation of the emission of
acceptor species j *via* their excitation spectra.
     From eqn. 29 it can be seen that the partitioning of
transferred energy between different acceptor species will
differ depending on the geometry in which they find themselves. This difference may be useful as an indicator of
which distribution model, 2-D or 3-D, is the more appropriate
in an unknown situation in which both models are plausible
but not unequivocally distinquished by a more direct interpretation of the data. By the same token, deviation of the
partitioning ratio from that predicted by either model may be
an indication of non-random local distribution of the acceptor
species about the donors.
     On rearranging eqn. 29 for the simplest case of a two-
acceptor system:

$$\phi_{Ti1}/\phi_{Ti} = \{1 + m_{i2}(R_{02}/R_{01})^i\}^{-1} \qquad (30)$$

Obviously, m_{i2} - the ratio of concentration of acceptor 2
to that of acceptor 1 - and the ratio of R_0 values to the
appropriate power, play inverse roles. In fact, on writing
eqn. 30 in the isomorphous form:

$$f = 1/(1 + ar^n) \qquad (31)$$

for simplicity, it can be seen by inspection that (nlog r + log a) is constant for a given partition of transferred energy f and eqn. 30 may be universally represented by the single curve of figure 2.

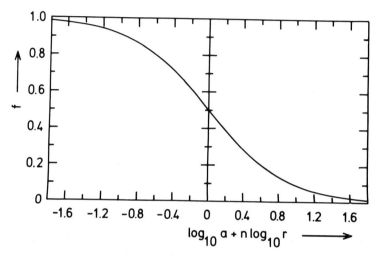

Fig. 2. Partitition of FRET efficiency between two competing acceptor species as represented by eqns. 30 and 31. The curve is valid for both n = 2 and 3 dimensions, a being the ratio of 2- or 3-D concentrations respectively of acceptor species 2 to species 1.

MATERIALS AND METHODS

A detailed description of the materials and methods used in this study will be given elsewhere (Novros *et al.*, in preparation). Brief summaries of the derivation of conA with fluorescein (donor) and tetramethylrhodamine (acceptor) isothiocyanates, their spectroscopic characterisation, the protocol for the cell surface energy-transfer experiments on yeast and cultured fibroblasts, and the determination of the amounts of derivatised conA's bound to the cells, are given here.

   The donor and acceptor thiocarbamyl adducts were formed by reacting fluorescein isothiocyanate or tetramethylrhodamine isothiocyanate with ε-lysine groups of commercial concan-avalin-A (FconA and TMRconA respectively) in a stabilising reaction medium (0.1M boric acid, 1M NaCl, 0.1M glucose and sub-mM concentrations of Ca^{2+} and Mn^{2+}) for several hours in the cold. After separation from the reaction medium by

Bio-gel filtration, the derivative was purified by affinity
chromatography on Sephadex, concentrated and stored in small
aliquots of a stabilising buffer at -20°C. The labelling
ratios were about 1 for FconA and about 0.6 for TMRconA,
based on tetradentate conA. Approximately 5% and 25%
respectively of the ligands appeared in the non-fluorescent
dimer form. In the latter case, the contributions of monomer
and dimer to the overall absorption spectrum were first
approximately estimated by matching the spectrum to those
obtained from a monomer-dimer equilibrium study (Selwyn and
Steinfeld, 1972) on a model adduct, TMR-aminocaproic acid.
This first-order match was then refined via the corrected
excitation spectrum for the fluorescent monomer. The latter
method alone sufficed to give a good enough estimate in the
case of FconA where only a small dimer contribution was
obtained.

All fluorescence measurements were made at excitation and
emission bandwidths of 4 nm on a Perkin-Elmer MPF-4 spectro-
fluorimeter with a depolariser in the excitation beam and
a polariser in the observation path oriented to obviate
polarisation bias in the recorded emission intensities
(Spencer and Weber, 1970). Excitation and emission spectra
were corrected for instrumental distortion by appropriate
calibration procedures using the rhodamine-B quantum counting
solutions provided and an Eastman Kodak white reflectance
standard. The excitation correction curve was checked with
dilute fluorescein and rhodamine-B solutions. For the highly
scattering cell suspensions it was necessary to determine
the excitation correction curves on the samples themselves,
having first established the monomer absorption spectra of
the conA derivatives, via their corrected excitation spectra
in optically clear dilute solutions. The donor quantum
yield, determined by comparison with fluorescein in 0.1M
NaOH (Weber and Teale, 1957), was of the order of one half.
R_0 values containing a refractive index of 1.333 and normalised to an orientation factor of unity were about 5.7 nm to
the monomer and 6.4 nm to the dimer acceptors on TMRconA.

For the energy-transfer experiments, four equal aliquots
of a suspension of yeast cells or fibroblasts were required.
In the former case, dried baker's yeast obtained commercially
was rehydrated in the appropriate buffer at 37°C and aliquots
of about 120×10^6 cells prepared as a wet centrifugate
BABL/c 3T3 murine fibroblasts, their simian virus transformed
derivative, SV3T3, and the spontaneously transformed variant
3T12 were cultured in Dulbecco's modified Eagle's medium
with 10% foetal calf serum and antibiotics at 37°C on Petri
dishes. After harvesting at confluence (two or three layers
growth in the case of transformed lines) with a trypsin-

collagenase-chick serum mixture, they were healed in suspension in growth medium under CO_2 for about half an hour, by
which time the intially increased conA binding capacity had
decreased to a plateau level. The cells were then washed
and distributed as above into aliquots of about $(10-30) \times 10^6$,
$(50-150) \times 10^6$ and 200×10^6 for 3T3, 3T12 and SV3T3 lines
respectively in order to obtain about the same order of
magnitude of fluorescence signal in all cases, including
the yeast. To the wet centrifugate were added appropriate
mixtures of native and derivatised conA's (at $10^{\circ}C$ for yeast
and $0^{\circ}C$ for fibroblasts) to a total of about 1.5 mg in 3 ml
per aliquot (about 30 fold more than is adsorbed onto the
cells). The cells were dispersed and kept in suspension by
gentle pumping with a Pasteur pipette. Initial clumping
was reversed early on by this procedure and the clumps did
not reform. After about an hour of incubation the cells were
thoroughly washed and resuspended in 4 ml buffer (at $10^{\circ}C$
for yeast, $2^{\circ}C$ for fibroblasts) for measurement. All
derivative-labelled cells at this stage showed classical
fluorescent ring-staining on examination under a fluorescence
microscope with appropriate filters. After the energy-
transfer measurements had been performed in the case of the
fibroblasts, the suspensions were incubated at $37^{\circ}C$ for about
an hour. Re-examination under the fluorescence microscope
revealed that the cells exhibited a combination of fluorescent patches and clusters with a few caps and some inter-
nalisation, but were not now ring-stained. After re-cooling
to $2^{\circ}C$, the energy-transfer measurements were repeated.

The four incubation mixtures for each experiment were made
up in buffer containing calcium and magnesium ions to ensure
full conA binding activity, and contained : (i) native conA
only; (ii) FconA and TMRconA at a ratio of about 1:2 to give
approximately equal absorbance at their respective absorption
peaks; (iii) FconA as in (ii) and native conA replacing the
acceptor derivative; (iv) TMRconA as in (ii) with native
conA replacing the donor derivative. Appropriate excitation
and emission spectra determined on the final cell suspensions
prepared as above from these incubation mixtures, in
combination with the absorption measurement described below,
in principle allows determination of energy transfer efficiencies by decrease of donor emission in presence of acceptor
(total transfer to both monomer and dimer acceptor fluorophors) and by the excitation spectrum of acceptor fluorescence (transfer to monomer acceptor only) corrected for background fluorescence and scatter in both cases, and for the
contribution of donor fluorescence in the latter case, as
well as providing the excitation spectral correction
indicated above.

After the energy-transfer experiments had been performed,
the cells were centrifuged down, the supernatant showing no
appreciable fluorescence. The yeast pellets were treated
with 30% sodium dodecyl sulphate at $70^{\circ}C$ to solubilise the
bound conA. After cooling and centrifuging off the remaining
cell fragments (not detectably fluorescent), the absorption
spectrum of the supernatants was used to obtain the amounts
of fluorophors bound to the cells, the sample originally
containing only native conA being used to correct for background absorption and scatter. In the case of the fibroblasts, the pellets were treated initially with small volumes
of 1% deoxycholate and 1 mg/ml DNAse solutions to obviate
the otherwise high viscosity of an extract made using only
sodium dodecyl sulphate. Since the absorption spectra of
the conA derivatives were altered considerably by such
treatment, the two extraction procedures were calibrated
directly with the original conA derivatives.

In the model yeast experiments it was found that acceptor
excitation spectra seemed to provide somewhat more consistent
measures of energy transfer efficiency than the donor
emission. This was attributed to the inherently lower error-
proneness of the former method. In this, the relative
concentrations of donor and acceptor (*via* the absorption
spectrum of the extract as detailed above) in a single
sample - sample (ii) - and the relative acceptor excitation
efficiency in the same sample primarily determine the energy
transfer value obtained. In the latter method, on the other
hand, donor concentrations and emission intensities are
determined on separate samples - (ii) and (iii) - for which,
since the relative binding efficiencies of the native and
two derivatised conA's are not necessarily equivalent, the
concentrations may differ inherently and between which, in
any case, the relative error will depend on the absolute
errors in determinations on two samples. The results
reported in the following section are thus based on the
former method. The acceptor excitation spectrum fully
corrected for background, interference from donor emission
at the acceptor observation wavelength, and for instrumental
excitation efficiency (including any effects arising from
the highly scattering nature of the sample) is compared with
the absorption spectrum pertaining to donor and acceptor
monomers in the transfer sample (sample (ii), equivalent to
the excitation spectrum that would be observed for 100%
efficient transfer to acceptor monomer), with that of the
acceptor monomer alone (equivalent to the excitation spectrum
for 0% transfer), and with those including intermediate
fractions of donor absorption between these levels
(corresponding to the excitation spectra which would be

observed for that fractional transfer efficiency).

Although the absorbance (A_{1cm}) of the donor and acceptor in the final suspension on which the FRET measurements were made was always kept low (< 0.03), it was possible that multiple scattering in the dense suspension might have resulted in a very long effective path length for emitted light. Any resulting 'trivial' energy transfer arising from reabsorption of donor emission by the acceptor would obviously upset the quantitative estimation of the efficiency of FRET. To check for occurrence of this 'trivial' process, approximately the same amounts of the conA derivatives as would have been bound to a yeast sample were added to the appropriate yeast suspensions in presence of 0.1M glucose to prevent binding. The transfer efficiency was measured in the usual way and turned out to be less than 0.02. The 'trivial' contribution to the measured transfer efficiency may therefore safely be neglected under the experimental conditions employed here.

The yeast cells were sized microscopically with a calibrated reticule under phase contrast, and approximated as ellipsoids of revolution to calculate the surface area. The thickness of the proteoglycan cell wall was taken to be 0.1 μm (Matile *et al.*, 1969). Estimates of the surface areas of the cultured fibroblasts were made from transmission electron micrographs of thin sections of cells to which native conA had been adsorbed according to the protocol given above for preparation of the energy transfer samples. The presence of both blebs and microvilli in these sections complicated the determination to the extent that the estimates approximate lower bounds for the surface area, but are probably within about ± 10% of the true values.

RESULTS AND DISCUSSION

A summary of the results obtained in the studies described above and a short discussion of their significance in relation to electron microscope data in the literature are given in the following. Detailed results will be presented elsewhere (Novros *et al.*, in preparation).

In the model experiments with yeast, the amount of conA adsorbed was, as expected, greatly in excess of that which could be accommodated as a monolayer on the surface : the square lattice separations for conA protomer would have been on the order of 1.5 nm whereas its diameter is almost three fold as great as this at about 4 nm, and similarly for dimer and tetramer. On the other hand, the corresponding cubic lattice separations throughout the proteoglycan cell wall were, at around 6 nm for monomer, entirely compatible with molecular dimensions, although implying a relatively uncondensed cell wall structure. With TMR monomer/conA protomer labelling ratios of between 0.1 and 0.16, energy transfer efficiencies of between 0.3 and 0.55 were observed. Such efficiencies would have required of the order of a 10-fold lower surface density of acceptor than would have obtained for the 2-D distribution. On the other hand the 3-D lattice separations obtained from these energy transfer results agreed quite satisfactorily with the amounts bound and the cell wall dimensions: the ratio of concentrations determined by energy transfer to those obtained from cell dimensions and amounts bound estimated from the absorption spectra of subsequently extracted conA, averaged 1.25 with a standard error of $\pm$ 0.27 over four separate determinations.

An example of the crude fluorescent data (actual noisy data smoothed for presentation) and the fully corrected excitation spectrum obtained from and compared with absorption spectra to obtain the transfer efficiency is given for one of the yeast samples in figure 3(a). The corresponding data for one of the fibroblast samples, exhibiting a much lower, but still well quantitated transfer efficiency is depicted in part (b) of the same figure.

The primary quantitative results and qualitative conclusions arising from the work on the normal and transformed fibroblast lines may be summarised as :

(i) There is little, if any, difference between the binding capacity per unit area of cell surface of the different fibroblast lines (quantitatively they differ by factors of up to 2 between lines and up to about 1.5 within a line over the small number of experiments carried out - 2 for 3T3 and SV3T3, 3 for 3T12). The discrepancy between this result and earlier work indicating about a 7 fold difference in density of bound conA between the parent 3T3 and derived SV3T3 lines (Collard and Temmink, 1975) may be attributable to the different regimes used to remove the cells from their culture substrates and heal them.

(ii) Within experimental error, there was no difference in energy transfer efficiencies between the ring-stained and patched conditions. The ratio of efficiencies under these conditions was (0.96 $\pm$ 0.08) averaged over 5 determinations. This infers that the local conA concentrations, that is, concentrations within regions of dimensions on the order of 30 nm (i.e. 0.03 µm or less which would not be resolved in the fluorescence microscope), are identical under these two conditions.

Fig. 3 (opposite) Examples of (smoothed) experimental excitation spectral FRET data for labelled conA adsorbed to yeast and fibroblasts (LHS) and of its fully reduced form displaying the quantitation of transfer efficiency to the fluorescent acceptor species (RHS). (a) results obtained on a yeast cell preparation. The designations (i)-(iv) refer to blank, donor and acceptor together, donor alone and acceptor alone respectively, as elaborated more fully in the text. The observation wavelength was 600 nm using excitation and emission bandwidths of 4 nm. (b) results from a BALB/c 3T12 preparation, designation and experimental conditions as for (a). Legend for RHS (———) theoretical excitation spectra for various energy transfer efficiencies ϕ_T from donor monomer to acceptor monomer. (—·——··——) examples of fully reduced experimental excitation spectra showing (a) 37% transfer efficiency (yeast) and (b) 15% transfer efficiency (3T12).

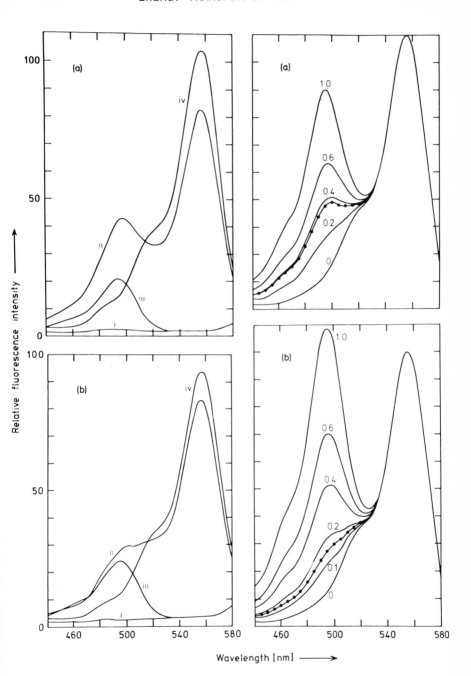

(iii)   No significance differences in transfer efficiencies
between the three cell types were noted : on average these
were (0.14 ± 0.01) for 3T3 (3 measurements) and 3T12 (6
measurements), and (0.14 ± 0.02) for SV3T3 (3 measurements).

There is thus no evidence adductable from the binding and
energy transfer experiments performed, for any differences
in the distribution of conA bound to the surfaces of the
trio of normal and malignant murine fibroblast lines under
the conditions described. In addition, even assuming a
random distribution of bound conA over the whole cell
surface area, the ratio between the 2-D concentrations
obtained from the amount bound to that required for the
energy transfer efficiency observed is (3.0 ± 0.3) over the
12 energy transfer measurements made. Obviously, this ratio
would increase by perhaps 5-fold or more on patching, and
the transfer efficiencies would be expected to go up to
between 60 and 90%, whereas in fact they do not change at all.
On the other hand, if an effectively three-dimensional
distribution is assumed, the amounts of conA bound, the cell
surface area estimates and the measured transfer efficiencies
are consistent with a random distribution of conA in a layer
at the surface of the cells whose thickness is estimated to
be of the order of 27.5 ± 6 nm (energy transfer range 14-
53 nm). Since the layer thickness thus determined is some
four to six-fold greater than the R_0 values for this system,
the equations for energy transfer in 3-D may still be
expected to apply to a very good approximation. If in fact
the conA-receptor complex is not entirely randomly distributed throughout such a volume, but tends to concentrate to
some extent within volumes of dimensions on the order of up
to a few times the R_0 value, the surface layer thickness
given will obviously represent an overestimate. The fact
that no increase in transfer efficiency is observed on
patching would still appear anomalous, but might be explained
by the piling up of this surface layer into thicker layers
without any necessary increase in the local 3-D concentration
of conA. This model for the conA-receptor complex distribution in ring-stained and patched conditions is essentially
consistent with the thin-section electron micrograph data
on conA bound to the plasma membranes of a number of cell
lines, using both ferritin-labelled (de Petris *et al.*, 1973;
de Petris, 1978) and unlabelled conA (de Petris, 1978). The
latter, in particular, reveals a 'relatively uniform dense
layer of 8-15 nm'(legend to figure 6 of de Petris, 1978)
at virtual saturation, while the piling up of such a layer
in the patched condition has also been visualised (de Petris,
1978).

CONCLUSIONS

Looked on as a feasibility study for the quantitative application of the FRET method to the study of distributions of cell surface components, the above results appear quite promising. Rather low energy transfer efficiencies can be measured quite accurately *via* acceptor excitation spectra (less so by decrease in donor emission) on very small quantities of fluorophors even in the presence of a dense and highly scattering cell suspension. Provided that the possible presence of transfer to non-emitting species (here the acceptor dimer) is properly taken into account, that the contribution of donor emission is taken into account, and that the amounts of both donor and acceptor bound can be accurately estimated, the energy transfer efficiency can be quantitatively interpreted with some degree of confidence.

Apart from the difficulties encountered with cell surface area estimation, the limiting factor in the present experiments appears to have been in quantitation of the amounts of material bound, and it was essentially these estimates which gave rise to the rather broad ranges in the parameters quoted in the previous section. On the other hand, over a number of determinations, the standard error obtained on those parameters indicate that they were nevertheless rather accurately determined. However, if such studies are to be extended to this and other lectin-acceptor complexes at non-saturating concentrations, as well as to specific antigen-antibody complexes, at very much lower overall concentrations at the cell surface (though they will only exhibit appreciable transfer efficiencies if bound at relatively high local concentration), it seems that radio-labelling of the fluorophors is likely to be the only satisfactory way to attain the required accuracy in quantitative bound donor and acceptor moieties. Further, if a high enough activity were attained, fully quantitative single cell determinations, both by fluorescence microscopy (Fernandez and Berlin, 1976) and cell-sorting cytofluorometry (Chan *et al.*, 1979) would be open to exploitation.

Finally, the use of steady-state measurements has the disadvantage that its basic assumptions are difficult, if not impossible, to check in systems as complex as those of interest here. The additional direct observation of the decay kinetics of both the donor and acceptor populations on the nanosecond time-scale when primarily exciting the donor (Fung and Stryer, 1978; Fernandez and Berlin, 1976) would go a long way towards checking these assumptions. The satisfactory resolution of changes in distribution of bound materials (Fernandez and Berlin, 1976) over relatively short times, i.e. on the order of tens of seconds to tens of minutes, by the nanosecond time-resolved FRET method is probably not feasible with conventional flashlamp excitation sources since their relatively low flash repetition rate necessitates long data collection times. However, the use of laser or synchrotron radiation sources with repetition rates in the several MHz range should certainly enable the eventual relatively routine realisation of such studies.

ACKNOWLEDGEMENTS

The authors wish to thank Dr. S. Ostrand-Rosenberg and Mr. J. Lye, for obtaining the electron micrographic data on which the cell surface area estimates were based.

REFERENCES

Birks, J.B. and Georghiou, S. (1968) *J. Phys. B.* $\underline{1}$, 958-65
Chan, S.S., Arndt-Jovin, D.J. and Jovin, T.M. (1979)
  J. Histochem. Cytochem. $\underline{27}$, 56-64
Collard, J.G. and Temmink, J.H.M. (1975) *J. Cell. Sci.*
  $\underline{19}$, 21-32
de Petris,S., Raff, M.C. and Mallucci, L. (1973)
  Nature New Biol. $\underline{244}$, 275-278
de Petris, S. (1978) Immunoelectron microscopy and immuno-
  fluorescence in membrane biology. *In* "Methods in Membrane
  Biology", Vol. 9. (E.D. Korn, ed.). Plenum Press, New
  York, pp. 1-201
Estep, T.N. and Thompson, T.E. (1979) *Biophys. J.* $\underline{26}$,
  195-207
Fernandez, S.M. and Berlin, R.D. (1976) *Nature* $\underline{264}$, 411-5
Forster, Th. (1949) *Z. Natur.* $\underline{4a}$, 321-327
Forster, Th. (1965). Delocalised excitation and excitation
  transfer.*In* "Modern Quantum Chemistry", pt. III.
  (O. Sinanoglu, ed.). Academic Press, New York, pp. 93-137
Fung, B.K-K. and Stryer, L. (1978) *Biochemistry* $\underline{17}$, 5241-48
Galanin, M.D. (1955) *Soviet-Physics-JETP* $\underline{1}$, 317-325
Hauser, M., Klein, U.K.A. and Gosele, U. (1976) *Z. Phys.*
  Chem. N.F., $\underline{101}$, 255-266
Koppel, D.E., Fleming, P.J. and Strittmatter, P. (1979)
  Biochemistry $\underline{18}$, 5450-5457
Maksimov, M.Z. and Rozman, I.M. (1962) *Opt. Spectrosc.*
  $\underline{12}$, 337-338
Matile, Ph., Moor, H. and Robincw, C.F. (1969) Yeast
  cytology.*In* "The Yeasts", Vol. 1. (A.H. Rose and J.S.
  Harrison, eds). Academic Press, New York, pp. 219-302
Nicolson, G.L. (1971) *Nature New Biol.*, $\underline{233}$, 244-246

Nicolson, G.L. (1972) *Nature New Biol.*, <u>239</u>, 193-197
Novros, J., Osrand-Rosenberg, S., Dale, R.E., Roth, S.,
  Edidin, M. and Brand, L. (1978) *Ber. Bunsenges Phys. Chem.*,
  <u>82</u>, 980
Schiller, P.W. (1975) The measurement of intramolecular
  distances by energy transfer. *In* "Biochemical Fluorescence:
  Concepts", (R.F. Chen and H. Edelhoch, eds). Marcel Dekker,
  New York, pp. 285-303
Selwyn, J.E. and Steinfeld, J.I. (1972) *J. Phys. Chem.*,
  <u>76</u>, 762-774
Shaklai, N.,Yguerabide, J. and Ranney, H.M. (1977a)
  Biochemistry, <u>16</u>, 5585-5592
Shaklai, N., Yguerabide, J. and Ranney, H.M.(1977b)
  Biochemistry, <u>16</u>, 5593-5597
Spencer, R.D. and Weber, G. (1970) *J. Chem. Phys.*,<u>52</u>, 1654-63
Steinberg, I.Z. (1968) *J. Chem. Phys.*, <u>48</u>, 2411-2413
Tweet, A.G., Bellamy, W.D. and Gaines, G.L. (1964) *J. Chem.
  Phys.*,<u>41</u>, 2068-2077
Weber, G. and Teale, F.W.J. (1957) *Trans. Faraday Soc.*,
  <u>53</u>, 646-655
Wolber, P.K. and Hudson, B.S. (1979) *Biophys. J.*, <u>28</u>,
  197-210

FLUORESCENT PROBES - A BIBLIOGRAPHY
(WITH AN EMPHASIS, FROM 1977 ONWARDS, ON
APPLICATIONS TO PROTEIN AND MEMBRANE STUDIES)

T. Tao, (1969), Biopolymers, 8, 609-32;
'Time-dependent fluorescence depolarisation and Brownian
rotational diffusion coefficients of macromolecules'

F.W.J. Teale and R.A. Badley, (1970), Biochem. J., 116, 341-8;
'Depolarisation of the intrinsic and extrinsic fluorescence
of pepsinogen and pepsin'

W.M. Vaughn and G. Weber, (1970), Biochemistry, 9, 464-73;
'Oxygen quenching of pyrenebutyric acid fluorescence in
water. Dynamic probe of the microenvironment'

R.F. Chen, (1970), Fluorescence News, 5, 1-2;
'Fluorescence depolarisation and lifetimes in macromolecular
chemistry'

J.K. Weltman and R.P. Davis, (1970), J. Mol. Biol., 54, 177-85;
'Fluorescence polarisation study of human IgA myeloma
protein : absence of segmental flexibility'

N. Okabe, (1970), Seibutsu Butsuri, 10, 151-61; Chem. Abstr.,
74-38240;
'Application of fluorescence depolarisation techniques to
studies of biopolymers, especially proteins'

W. Terpstra, (1970), Biochim. Biophys. Acta, 216, 179-91;
'Particle fractionation from spinach leaf homogenates and
diffusion of spinach protein factor'

B. Witholt and L. Brand, (1970), Biochemistry, 9, 1948-56;
'Multioscillator fluorescence depolarisation, anisotropy of
dye binding'

R.F. McGuire, (1970), Diss. Abstr. Int. B, 30, 3042-2;
'Fluorescence and polarisation of fluorescence of protein
solutions'

J.R. Brocklehurst, R.B. Freedman, D.J. Hancock and G.K. Radda, (1970), Biochem. J., 116, 721-31;
'Membrane studied with polarity-dependent and excimer-forming fluorescent probes'

G. Weber, M. Shinitzky, A.C. Dianoux and C. Gitler, (1971), Biochemistry, 10, 2106-13;
'Microviscosity and order in the hydrocarbon region of micelles and membranes determined with fluorescent probes. 1. Synthetic micelles'

T. Tao, (1971), Biochem. J., 122, 54;
'Rotational mobility of a fluorescent rhodopsin derivative in the rod outer-segment membrane'

L. Brand and J.R. Gohlke, (1971), J. Biol. Chem., 246, 2317-19;
'Nanosecond time-resolved fluorescence spectra of a protein-dye complex'

R.A. Badley, H. Schmeider and W.G. Martin, (1971), Biochem. Biophys. Res. Commun., 45, 174-83;
'Orientation and motion of a fluorescent probe in model membranes'

J.F. Faucon and C. Lussan, (1971), C.R. Acad. Sci., Ser. D, 273, 646-9;
'Polarisation of the fluorescence of 8-anilino-1-napthalene-sulphonate (ANS) bound in the phospholipid vesicular liposomes'

P. Wahl, M. Kasai and J-P. Changeux, (1971), Eur. J. Biochem., 18, 332-341;
'A study on the motion of proteins in excitable membrane fragments by nanosecond fluorescence polarisation spectroscopy'

G. Weber and D.P. Borris, (1971), Mol. Pharmacology, 7, 530-537;
'The use of a cholinergic fluorescent probe for the study of the receptor proteolipid'

C.T. Ladoulis and T.J. Gill, III, (1971, pub. 1972), Protides Biol. Fluids, Proc. Colloq., 19, 403-7;
'Studies of polypeptide structure by fluorescent determination of fluorescence lifetimes'

T.J. Gill, III, C.T. Ladoulis, M.F. King and H.W. Kunz,
(1971, pub. 1972), Protides Biol. Fluids, Proc. Colloq.,
19, 383-92;
'Studies of polypeptide structure by fluorescence techniques.
7. Use of fluorescent techniques for studying the structures
and structural transitions of polypeptides in solution'

D.K.M. Young, (1972), Diss. Abstr. Int. B, 33, 2498;
'Metal ion-induced changes in the structure of proteins
investigated with fluorescence and fluorescence polarisation
techniques'

H.P. Zingsheim and D.A. Haydon, (1972), Naturwissenschaften,
59, 419-20;
'Fluorescence spectroscopy of planar black lipid membranes.
Probe adsorption and quantum yield determination'

A.B. Rawitch, (1972), Arch. Biochem. Biophys., 151, 22-7;
'Rotational diffusion of bovine α-lactalbumin. Comparison
with egg white lysozyme'

R.T. Mayer and C.M. Himel, (1972), Biochemistry, 11, 2082-90;
'Dynamics of fluorescent probe-cholinesterase reactions'

P.L. Coleman, H. Weiner, A.M. Kaplan and M.J. Freeman, (1972),
J. Immunol. Methods, 1, 145-53;
'Application of fluorescence emission and polarisation for
the measurement of association constants between fluorescein-
conjugated protein antigens and antibodies'

J. Schlessinger and I.Z. Steinberg, (1972), Proc. Nat. Acad.
Sci. U.S., 69, 769-72;
'Circular polarisation of fluorescence of probes bound to
chymotrypsin. Change in asymmetric environment upon
electronic excitation'

C.G. Dos Remedios, R.G.C. Millikan and M.F. Morales, (1972),
J. Gen. Physiol., 59, 103-20;
'Polarisation of tryptophan fluorescence from single
striated muscle fibres. Molecular probe of contractile state'

C. Huang and J.P. Charlton, (1972), Biochem. Biophys. Res.
Commun., 46, 1660-6;
'Studies on the state of phosphatidylcholine molecules before
and after ultrasonic and gel-filtration treatments'

S.J. Singer and G.L. Nicolson, (1972), Science, 175,720-731;
'The fluid mosaic model of the structure of cell membranes'

R.W. Barker, J.D. Bell, G.K. Radda and R.E. Richards, (1972),
Biochim. Biophys. Acta, 260, 161-63;
'Phosphorus nuclear magnetic resonance in phospholipid
dispersions'

G. Weber, Ann. Rev. Biophys. Bioengineering, (1972), 1, 553-
569;
'Uses of fluorescence in biophysics : some recent developments'

J.F. Faucon and C. Lussan, (1973), Biochim. Biophys. Acta,
307, 459-66;
'Aliphatic chain transitions of phospholipid vesicles and
phospholipid dispersions determined by polarisation of
fluorescence'

K.A. Jacobson, (1973), Diss. Abstr. Int. B, 33, 3501;
'Fluorescent probe rotation within lipid bilayer membranes'

H.P. Zingsheim and D.A. Haydon, (1973), Biochim. Biophys.
Acta, 298, 755-768;
'Fluorescence spectroscopy of planar black lipid membranes.
Probe absorption and quantum yield determinations'

J. Yguerabide, (1972, pub. 1973), Fluoresc. Tech. Cell. Biol.,
367-79; Chem. Abstr., 83-160057;
'Conformational dynamics of model membranes'

J. Yguerabide, (1972, pub. 1973), Fluoresc. Tech. Cell. Biol.,
311-31; Chem. Abstr., 83-160055;
'Nanosecond fluorescence spectroscopy of biological macromolecules and membranes'

B. Myhr and J. Foss, (1972, pub. 1973), Fluoresc. Tech. Cell.
Biol., 291-9; Chem. Abstr., 83-160215;
'Fluorescence, absorption and optical rotatory behaviour of
acridine orange - poly-L-glutamic acid complexes'

B. Witholt and L. Brand, (1972, pub. 1973), Fluoresc. Tech.
Cell Biol., 283-90; Chem. Abstr., 83-174839;
'Effect of the excitation wavelength on the fluorescence
depolarisation of protein-dye complexes or conjugates'

U. Cogan, M. Shinitzky, G. Weber and T. Nishida, (1973),
Biochemistry, 12, 521-8;

'Microviscosity and order in the hydrocarbon region of
phospholipid and phospholipid-cholesterol dispersions
determined with fluorescent probes'

P.W. Schiller and R. Schwyzer, (1971, pub. 1973), Proc.
Eur. Pept. Sym., 11th, 354-366; Chem. Abstr., 84-44623;
'21-N-ε-dansyllysine-corticotropin-(1,24)-tetracosipeptide,
a biologically active derivative of ACTH. Receptor protein
and studies with fluorescence depolarisation and intramolecular excitation energy transfer'

R.A. Badley, W.G. Martin and H. Schneider, (1973),
Biochemistry, 12, 268-75;
'Dynamic behaviour of fluorescent probes in lipid bilayer
model membranes'

J.R. Lakowicz and G. Weber, (1973), Biochemistry, 12, 4171-9;
'Quenching of protein fluorescence by oxygen. Detection of
structural fluctuations in proteins on the nanosecond time
scale'

J.H. Easter and L. Brand, (1973), Biochem. Biophys. Res.
Comm., 52, 1085-92;
'Nanosecond time-resolved emission spectroscopy of a
fluorescence probe bound to egg L-α-lecithin vesicles'

D. Papahadjopoulos, K. Jaconson, S. Nir and T. Isac, (1973)
Biochim. Biophys. Acta, 311, 330-48;
'Phase transitions in phospholipid vesicles. Fluorescence
polarisation and permeability measurements concerning the
effect of temperature and cholesterol'

J. Hawiger and S. Timmons, (1973), Biochem. Biophys. Res.
Comm., 55, 1278-84;
'Dynamic changes in the membrane of leukocytic lysosomes
detected with fluorescent probe and accompanied by release
of lysosomal enzymes'

E.V. Anufrieva, Y.Y. Gotlib, M.G. Krakovyak, I.A. Torchinskii,
T.V. Sheveleva and B.V. Shestopalov, (1973), Vysokomol.
Soedin, ser. A, 15, 2538-48; Chem. Abstr., 80-96475;
'Molecular-weight dependence of the rotational mobility of
mamcromolecules in solution studied by a polarised lumin-
escence method'

P. Lipton, (1973), Biochem. J., 136, 99-109;
'Effects of membrane depolarisation of nicotinamide nucleotide fluorescence in brain slices'

M. Inbar, M. Shinitzky and L. Sachs, (1973), J. Mol. Biol.,
81, 245-53;
'Rotational relaxation time on concanavalin A bound to the
surface membrane of normal and malignant transformed cells'

J.F. Faucon and C. Lussan, (1973), C.R. Acad. Sci., Ser. C,
277, 591-4;
'Fluorescence polarisation study of the effect of oxidation
on lamellar lipid structure'

M. Shinitzky and Y. Barenholz, (1974), J. Biol. Chem., 249,
2652-2657;
'Dynamics of the hydrocarbon layer in liposomes of lecithin
and sphingomyelin containing dicetylphosphate'

D. Genest, Ph. Wahl and J.C. Auchet, (1974), Biophys. Chem.,
1, 266-278;
'The fluorescence anisotropy decay due to energy transfers
occurring in the ethidium bromide-DNA complex. Determination of the deformation angle of the DNA helix'

S.W. Hui, D.F. Parsons and M. Cowden, (1974), Proc. Nat.
Acad. Sci. U.S.A., 71, 5068-72;
'Electron diffraction of wet phospholipid bilayers'

T. Alvaeger and W.X. Balcavage, (1974), Biochem. Biophys.
Res. Comm., 58, 1039-46;
'Nanosecond fluorescence decay study of mitochondria and
mitochondrial membranes'

D.H. Hayes and B.C. Pressman, (1974), J. Membrane Biol.,
16, 195-205;
'Calcium^{2+} ion ionophore with a polarity-dependent and
complexation-dependent fluorescence signal'

K. Jaconson and D. Wobschall, (1974), Chem. Phys. Lipids,
12, 117-31;
'Rotation of fluorescent probes localised within lipid
bilayer membranes'

S. Nakai and C.M. Kason, (1974), Biochim. Biophys. Acta,
351, 21-7;
'Fluorescence study of the interactions between κ- and α-SI-
casein and between lysozyme and ovalbumin'

L.J. Andrews and L.S. Forster, (1974), Photochem. Photobiol.
19, 353-60;

'Fluorescence characteristics of indoles in nonpolar
solvents. Lifetimes, quantum yields and polarisation spectra'

E.P. Bakker and K. Van Dam, (1974), Biochim. Biophys. Acta,
339, 157-63;
'Influence of diffusion potentials across liposomal membranes
on the fluorescence intensity of 1-anilinonaphthalene-8
sulphonate'

S. Cheng, J.K. Thomas and C.F. Kulpa, (1974), Biochemistry,
13, 1135-9;
'Dynamics of pyrene fluorescence in E. Coli membrane
vesicles'

C. Lussan and J.F. Faucon, (1974), Biochim. Biophys. Acta,
345, 83-90;
'Effects of ions on vesicles and phospholipid disperions
studied by polarisation of fluorescence'

M. Shinitzky, (1974), Isr. J. Chem., 12, 879-90;
'Fluidity and order in the hydrocarbon-water interface of
synthetic and biological micelles as determined by fluorescence polarisation'

D.C.H. Yang, W.E. Gall and G.M. Edelman, (1974), J. Biol.
Chem., 294, 7018-23;
'Rotational correlation time of concanavalin A after
interaction with a fluorescent probe'

R. Peters, J. Peters, K.H. Tews and W. Bachr, (1974)
Biochim. Biophys. Acta, 367, 282-94;
'Microfluorimetric study of translational diffusion in
erythrocyte membranes'

Y.S. Borovikov, N.A. Chernogryadskaya and Y.M. Rozanov,
(1974), Tsitologiya, 16, 977-82; Chem. Abstr. 81-148870;
'Structural changes in myosin and actin filaments in muscle
fibre studied by the polarised ultraviolet fluorescent
microscopy method'

J.M. Vanderkooi and J.B. Callis, (1974), Biochemistry, 13,
4000-6;
'Pyrene. Probe of lateral diffusion in the hydrophobic
region of membranes'

M.S. Tung and R.F. Steiner, (1974), Biochem. Biophys. Res.
Comm., 57, 876-86;

'Application of nanosecond fluorometry to an allosteric protein'

I.H. Leaver and D.E. Rivett, (1974), Mol. Photochem., 6, 113;
'Structural and solvent effects on the fluorescence of 1,3-diphenyl-2-pyrazolines'

K. Mihashi and Y. Kawasaki, (1974), Tampakushitu Kakusan Koso, Bessatsu, 198-205; Chem. Abstr., 81-121016;
'Micro-Brownian motion of proteins observed by the fluorescence polarisation method'

J. Vanderkooi, S. Fischkoff, B. Chance and R.A. Cooper (1974), Biochemistry, 13, 1589-95;
'Fluorescent probe analysis of the lipid architecture of natural and experimental cholesterol-rich membranes'

J. Vanderkooi, J. Callis and B. Chance, (1974), Histochem. J., 6, 301-310;
'Use of the fluorescent dye, pyrene, to study the dynamic aspects of membrane structure'

S. Cheng, M. Graetzel, J.K. Thomas and C.F. Kulpa (1973 pub. 1975), Fast Processes Radiat. Chem. Biol., Proc. L.H. Gray Conf., 5th, 193-211; Chem. Abstr., 84-176152;
'Dynamics of pyrene fluorescence in micelles and membrane vesicles'

S. Schuldiner, R.D. Spencer, G. Weber, R. Weil and H.R. Kaback (1975), J. Biol. Chem., 250, 8893-6;
'Mechanisms of active transport in isolated bacterial membrane vesicles. XXX. Lifetime and rotational relaxation time of dansylgalactoside bound to the lac carrier protein'

A. Bigi, G. Curatola, L. Mazzanti and G. Lenaz, (1975), Boll. Soc. Ital. Biol. Sper., 51, 298-301;
'General anesthetics and biomembrane fluidity. II Fluorescent probe studies'

P. Fuchs, A. Parola, P.W. Robbins and E.R. Blout, (1975), Proc. Natl. Acad. Sci. U.S.A., 72, 3351-4;
'Fluorescence polarisation and viscosities of membrane lipids of 3T3 cells'

M.R. Eftink and C.A. Ghiron, (1975), Proc. Natl. Acad. Sci. U.S.A., 72, 3290-4;
'Dynamics of a protein matrix revealed by fluorescence quenching'

R. McGuire and I. Feldman, (1975), Polymers, 14, 1095-1102;
'Static and dynamic quenching of protein fluorescence. II.
Lysosyme'

E. Frehland and H.W. Trissl, (1975), J. Membr. Biol., 21,
147-173;
'Fluorescence polarisation in a planar array of pigment
molecules. Theoretical treatment and application to flavins
incorporated into artificial membranes'

M.P. Blaustein and J.M. Goldring, (1975), J. Physiol.(London),
247, 589-615;
'Membrane potentials in pinched-off presynaptic nerve
terminals monitored with a fluorescent probe. Evidence that
synaptosomes have potassium diffusion potentials'

I. Feldman, D. Young and R. McGuire, (1975), Biopolymers,
14, 335-51;
'Static and dynamic quenching of protein fluorescence. I.
Bovine serum albumin'

J.M. Vanderkooi, S. Fischkoff, M. Andrich, F. Podo and
C.S. Owen, (1975), J. Chem. Phys., 63, 3661-6;
'Diffusion in two dimensions. Comparison between diffusional
fluorescence quenching in phospholipid vesicles and in
isotropic solution'

A. Jonas and R.W. Jung, (1975), Biochem. Biophys. Res. Comm.,
66, 651-7;
'Fluidity of the lipid phase of bovine serum. High density
lipoprotein from fluorescence polarisation measurements'

Ph. Wahl, (1975), Chem. Phys., 7, 210-219;
'Fluorescence anisotropy of chromophores rotating between
two reflecting barriers'

R.J. Cherry, (1975), FEBS Letters, 55, 1-7;
'Protein mobility in membranes'

F.J. Barrantes, B. Sakmann, Ro. Bonner, H. Eibl and
T.M. Jovin, (1975), Proc. Natl. Acad. Sci. U.S.A., 72,
3097-3101;
'1-pyrene-butyrylcholine : a fluorescent probe for the
cholinergic system'

C.W. Wu, L.R. Yarborough, Z. Hillei and F.Y-H. Wu, (1975)
Proc. Natl. Acad. Sci. U.S.A., 72, 3019-3023;
'Signa cycle during in vitro transcription : demonstration
by nanosecond fluorescence depolarisation spectroscopy'

L.W. Harrison, D.S. Auld and B.L. Vallee, (1975), Proc.
Natl. Acad. Sci. U.S.A, 72, 4356-4360;
'Intramolecular arsanilazotyrosine-248-Zn complex of carboxy-
peptidase A : a monitor of multiple conformational states
in solution'

P. Strittmatter and M.J. Rogers, (1975), Proc. Natl. Acad.
Sci. U.S.A., 72, 2658-2661;
'Apparent dependence of interactions between cytochrome b5
and cytochrome b5 reductase upon translational diffusion
in dimyristoyl lecithin liposomes'

P. Chakrabarti and H.G. Khorana, (1975), Biochemistry, 14,
5021-5032;
'A new approach to the study of phospholipid-protein interactions in biological membranes. Synthesis of fatty acids
and phospholipids containing photosensitive groups'

E. Haas, M. Wilchek, E. Katchalski-Katzir and I.Z. Steinberg,
(1975), Proc. Natl. Acad. Sci. U.S.A., 72, 1807-1811;
'Distribution of end-to-end distances of oligopeptides in
solution as estimated by energy transfer'

W.G. Pohl and J. Teissie, (1975), Z. Naturforsch., 30,
147-151;
'The use of fluorescent probes for studying the interaction
of proteins with black lipid membranes'

B. Chance and M. Baltscheffsky, (1975), Biomembranes, 7,
33-55;
'Carotenoid and merocyanine probes in chromatophore membranes

J.C. Metcalfe, DHEW Publ. (N.I.H., USA), (1975), (Cell Surf.
Malig. Workshop), 21-33 ; Chem. Abstr., 87-49512;
'Physical probes of plasma membrane structure'

G.E. Dobretsov and Y.A. Vladimorov, (1975), Usp. Biol. Khim.,
16, 115-34; Chem. Abstr., 87-97491;
'Study of proteins and membranes by using fluorescent probes'

J. Lenard, F.R. Landsberger, C.Y. Wong, P.W. Choppin and
R.W. Compans, (1975), Negat. Strand Viruses, pap. symp.

823-33, Chem. Abstr. 87-148360;
'Organisation of lipid and protein in viral membranes : spin
label and fluorescence probe studies'

H.H. Gruenhagen and J.P. Changeux, (1976), J. Mol. Biol.,
106, 497-516;
'Studies on the electrogenic action of acetylcholine with
torpedo marmorata electric organ. IV. Quinacridine : a
fluorescent probe with the conformational transitions of the
cholinergic receptor protein in its membrane-bound state'

H.E. Edwards, J.K. Thomas, G.R. Burleson and C.F. Kulpa,
(1976), Biochim. Biophys. Acta, 448, 451-9;
'The study of Rous sarcoma virus-transformed baby hamster
kidney cells using fluorescent probes'

J. Teissie, J.F. Tocanne and A. Baudras, (1976), FEBS Lett.,
70, 123-6;
'Phase transitions in phospholipid monolayers at the air-
water interface : a fluorescence study'

J.C. Smith, P. Russ, B.S. Cooperman and B. Chance, (1976),
Biochemistry, 15, 5094-105;
'Synthesis, structure determination, spectral properties,
and energy-linked spectral responses of the extrinsic probe
oxonal U in membranes'

E. Sackmann, (1976), Z. Phys. Chem. (Frankfurt), 101, 391-
416;
'On the application of excimers as optical probes in membrane
research'

W.A. Cramer, P.W. Postma and S.L. Helgerson, (1976),
Biochim. Biophys. Acta, 449, 401-411;
'An evaluation of N-phenyl-1-naphthylamine as a probe of
membrane energy state in E.Coli'

W.E. Mueller, (1976), Hoppe-Seyler's Z. Physiol. Chem., 357,
1487-94; Chem. Abstr., 86-38845;
'The use of 8-anilino-1-naphthalenesulphonic acid as a
reporter group molecule for circular dichroism and fluorescence measurements. The effect of stearic acid and sodium
dodecyl sulphate on the conformation of bovine and human
serum albumin'

D.L. Vandermeulen and Govindjee, (1976), Biochim. Biophys.
Acta, 449, 340-56;
'Anthroyl stearate as a fluorescent probe of chloroplast
membranes'

M. Kapoor, (1976), Biochem. Rev., 47, 27-51;
'The application of fluorescence and fluorescent probes in
biological systems'

Y. Katsumata, (1976), Nagoya Igaku, 99, 27-33; Chem. Abstr.,
86-116388;
'Studies on membrane components of sarcoplasmic reticulum
using a fluorescent probe'

A.E. Zakaryan, A.B. Avakyan and G.A. Panosyan, (1976), Biol.
Zh. Arm., 29, 65-9; Chem. Abst., 86-151812;
'Interaction of plasma membrane fractions with dyes Eosine
BA and Rhodamine 6Zh'

A. Waggoner, (1976), Enzymes Biol. Membr., 1, 119-37, Plenum
Press, N.Y., Chem. Abstr., 86-152180;
'Fluorescent probes of membranes'

W.T. Schaffer and M.L. Olson, (1976), J. Neurochem., 27,
1319-25,
'Chlorotetracycline-associated fluorescence changes during
calcium uptake and release by rat brain synaptosomes'

T. Skine and T. Ohyashiki, (1976), Kagaku No Ryoiki, Zokan,
114, 31-7; Chem. Abstr., 87-1901;
'Design and synthesis of thiol-directed fluorescent probes
and their application to biological systems'

J. Vanderkooi and A. McLaughlin, (1976), Biochem. Fluoresc.
Concepts, 2, 737-65, Dekker, New York; Chem. Abstr., 87-
148068;
'Use of fluorescent probes in the study of membrane structure
and function'

R. Kraayenhof, J.R. Brocklehurst and C-P. Lee, (1976),
Biochem. Fluoresc. Concepts, 2, 767-809, Dekker, New York;
Chem. Abstr., 87-148069;
'Fluorescent probes for the energised state in biological
membranes'

J.E. Churchill, (1976), Mod. Fluoresc. Spectrosc., 2, 217-37,
Plenum Press, New York;
'Fluorescent probe studies of binding sites in proteins and
enzymes'

C.L. Bashford, C.G. Morgan and G.K. Radda (1976), Biochim.
Biophys. Acta, 426, 157-72;

J.I. Morgan, J.S. Bramhall, A.Z. Britten and A.D. Perris, (1976), J. Endocrinol., 69, 29P-30P;
'The use of a fluorescent probe to demonstrate the inter-action of calcium and estradiol-17 in the thymic lymphocyte plasma membrane'

I. Iweibo,(1976), Biochim. Biophys. Acta, 446, 192-205;
'Protein fluorescence and electronic energy transfer in the determination of molecular dimensions and rotational relaxation times of native and coenzyme-bound horse liver alcohol hydrogenase'

J.H. Easter, R.P. DeToma and L. Brand, (1976), Biophys. J., 16, 571-83;
'Nanosecond time-resolved emission spectroscopy of a fluorescent probe adsorbed to L-α-egg lecithin vesicles'

P.A. George Fortes, (1976), Mitochondria: Bioenerg. Biog. Membr. Struct., 327-348; Chem. Abstr., 85-15859;
'Nanosecond fluorescence spectroscopy of biological membranes'

J.C. Brochon, Ph. Wahl, P. Vachette and M.P. Daun, (1976), Eur. J. Biochem., 65, 35-9;
'Nanosecond-pulse fluorometry study of S4 ribosomal protein'

R.R. Alfano, W. Yu, R. Govindjee, B. Becher and T.G. Ebrey, (1976), Biophys. J., 16, 541-5;
'Picosecond kinetics of the fluorescence from the chromo-phore of the purple membrane protein of Halobacterium Halobium'

A. Spisni, A.M. Sechi, P. Guadagnini and L. Masotti, (1976), Boll. Soc. Ital. Biol. Sper., 52, 487-492 ; Chem. Abstr., 85-138913;
'Effects of polyamines on mitichondrial membranes : studies using fluorescent probes'

R.P. DeToma, J.H. Easter and L. Brand,(1976), J. Amer. Chem. Soc., 98, 5001-7;
'Dynamic interactions of fluorescent probes with the solvent environment'

E. Weidekamm, C. Schudt and D. Brdiczka, (1976), Biochim. Biophys. Acta, 443, 169-180;
'Physical properties of muscle cell membranes during fusion. A fluorescence polarisation study with the ionophore A23187'

N.F. Moore, Y. Barenholz and R.R. Wagner, (1976),
J. Virol., 19, 126-35;
'Microviscosity of togavirus membranes studied by fluorescence depolarisation. Influence of envelope proteins
and the host cell'

M. Kopelman, U. Cogan, S. Mokady and M. Shinitzky, (1976),
Biochim. Biophys. Acta, 439, 449-60;
'The interaction between retinol-binding proteins and
prealbumins studied by fluorescence polarisation'

G.W. Stubbs, B.J. Litman and Y. Barenholz, (1976),
Biochemistry, 15, 2766-72;
'Microviscosity of the hydrocarbon region of the bovine
retinal rod outer segment disc membrane determined by
fluorescent probe measurements'

J.F. Faucon, J. Dufourcq, C. Lussan and R. Bernon, (1976),
Biochim. Biophys. Acta, 436, 283-94;
'Lipid protein interactions in membrane models. Fluorescence
polarisation study of cytochrome B5-phospholipid complexes'

L-H. Tang, Y. Kubota and R.F. Steiner, (1976), Biophys.
Chem., 4, 203-213;
'A comparative study of bovine, α-lactalbumin and lysozyme
by nanosecond fluorometry'

B.R. Lentz, Y. Barenholz and T.E. Thompson, (1976),
Biochemistry, 15, 4521-8;
'Fluorescence depolarisation studies of phase transitions
and fluidity in phospholipid bilayers : 1.Single component
phosphatidylcholine liposomes'

B.D. Goldstein, G.W. Falk, L.J. Benjeman and E.M. McDonagh,
(1976), Blood Cells, 2, 535-40;
'Alteration in the chloroform quenching of red cell membrane
native protein fluorescence following incubation with
malonaldehyde and other crosslinking agents'

Y.S. Borovikov, N.A. Chernogryadskaya, Y.M. Rozanov,
M.S. Shudel and V.I. Stabrovskaya, (1976), Biofiz. Biokhim.
Myshechnogo Sokrashcheniya, 145-50; Chem. Abstr., 86-136730;
'Ultraviolet-fluorescence polarisation microscopic study
of the characteristics of the microstructure and conform-
ational changes of muscle fibre contractile proteins'

E. Carbone, F. Malerba and M. Poli, (1976), Biophys.
Struct. Mech., 2, 251-66; Chem. Abstr., 86-102252;
'Orientation and rotational freedom of fluorescent probes
in lecithin bilayers'

Y.S. Borovikov, M.S. Bogdanova, N.A. Chernogryadskaya,
Y.M. Rozanov and V.P. Kirillina, (1976), Tsitologiya, 18,
1502-5; Chem. Abstr., 86-53285;
'Polarised ultraviolet fluorescence microscopic study of
structural changes in muscle fibre contractile proteins.
II. Effect of the troponin-tropomyosin complex on F-actin
conformation in muscle fibre'

D. Axelrod, P. Ravdin, D.E. Koppel, J. Schlessinger, W.W.
Webb, E.L. Elson and T.R. Podleski, (1976), Proc. Natl.
Acad. Sci. U.S.A., 73, 4594-8;
'Lateral motion of fluorescently-labelled acetylcholine
receptors in membranes of developing muscle fibres'

S.R. Aragon San Juan, (1976), Diss. Abstr. Int. B, 37,
2266;
'Part I. Dynamic light scattering from macromolecules with
applications to polydisperse systems. Part II. Fluorescence
correlation spectroscopy as a probe of molecular dynamics'

H. Rubsamen, P. Barald and T. Podleski, (1976), Biochim.
Biophys. Acta, 455, 767-9;
'A specific decrease of the fluorescence depolarisation of
perylene in muscle membranes from mice with muscular
dystrophy'

D.E. Koppel, D. Axelrod, J. Schlessinger, E.L. Elson and
W.W. Webb, (1976), Biophys. J., 16, 1315-29;
'Dynamics of fluorescence marker concentration as a probe
of mobility'

Y.S. Borovikov, N.A. Chernogryadskaya, M.S. Bogdanova,
Y.M. Rozanov and V.P. Kirillina, (1976), Tsitologiya,
18, 1371-7; Chem. Abstr., 86-27051;
'Polarised ultraviolet fluorescent microscopic study of the
structural changes of muscle fibre contractile proteins.
I. Conformational changes in the F-actin in muscle fibre
induced by ATP and its analogues'

G. Weber, S.L. Helgerson, W.A. Cramer and G.W. Mitchell,
(1976), Biochemistry, 15, 4429-32;
'Changes in the rotational motion of a cell-bound fluorophor
caused by colicin E1 : a study by fluorescence polarisation

and differential polarised phase fluorometry'

J.F. Faucon, J.J. Piaud and C. Lussan, (1976), J. Chim.
Phys. Phys.-Chim. Biol., 73, 658-64;
'Construction of an apparatus for accurate measurements of
the polarisation of fluorescence : example of application
to models of biological membranes'

D. Schachter, U. Cogan and M. Shinitzky, (1976), Biochim.
Biophys. Acta, 448, 620-4;
'Interaction of retinol and intestinal microvillus membranes
studied by fluorescence polarisation'

B.R. Lentz, Y. Barenholz and T.E. Thompson, (1976),
Biochemistry, 15, 4529-37;
'Fluorescence polarisation studies of phase transitions and
fluidity in phospholipid bilayers. 2. Two-component phosph-
atidycholine liposomes'

R.A. Badley, (1976), Mod. Fluoresc. Spectrosc., 2, 91-168;
Chem. Abstr., 88-18397;
'Fluorescent probing of dynamic and molecular organisation
of biological membranes'

Y. Kawasaki, N. Wakayama and T. Oshima, (1976), Kagaku No
Ryoiki Zokan, 114, 39-46; Chem. Abstr., 87-1900;
'Studies on the dynamic properties of biological substances
using a nanosecond fluorescence photometer'

R. Narayanan and P. Balaram, (1976), Biochem. Biophys. Res.
Comm., 70, 1122-1127;
'Synthesis and fluorescence properties of a probe for
membrane anionic sites'

T. Vo-Dinh, K.P. Li and J.D. Winefordner, (1976), Biochem.
Biophys. Res. Comm., 73, 187-191;
'Fluorescence studies of benzo-(a)-pyrene in liposome
membrane systems'

A.E. McGrath, C.G. Morgan and G.K. Radda, (1976), Biochim.
Biophys. Acta, 426, 173-185;
'Photobleaching. A novel fluorescence method for diffusion
studied in lipid systems'

M.P. Andrich and J.M. Vanderkooi, (1976), Biochemistry,
15, 1257-1261;
'Temperature dependence of 1,6-diphenyl-1,3,5-hexatriene
fluorescence in phospholipid artificial membranes'

'Measurement and interpretation of fluorescence polarisations in phospholipid dispersions'

A. Spisni, A.M. Sechi and L. Mosotti, (1976), Boll. Soc. Ital. Biol. Sper., 52, 493-6; Chem. Abstr., 85-172205;
'Interactions of polyamines with mitochondrial membranes. Study with fluorescence polarisation'

U. Pick and M. Avron, (1976), Biochim. Biophys. Acta, 440, 189-204;
'Measurement of transmembrane potentials in Rhodospirillum Rubrum chromatophores with an oxacarbocyanine dye'

A. Romero, J. Sunamoto and J.H. Fendler, (1976), Colloid Interface Sci., 5, 111-18; Chem. Abstr., 87-80072;
'Steady state and nanosecond time resolved fluorescence of dansyl N-octadecyl amine in bilayer liposomes'

E.A. Permyakov, (1977), Ravnovesnaya Din. Nativnoi Strukl. Belka, 83-99; Chem. Abstr. 91-15511;
'Cooperative freezing of the intramolecular mobility of proteins'

A. Jonas, S.M. Drengler and A.M. Scanu, (1977), Biochem. Biophys. Res. Comm., 78, 1424-30;
'Fluorescence polarisation studies of human and rhesus A-1 apolipoproteins and their complexes with phosphatidylcholine'

S. Kawato, A. Ikegami and K. Kinoshita, (1977), Hyomen, 15, 529-38;   Chem. Abstr., 87-180037;
'Method for studying active characteristics of membrane lipids and proteins by nanosecond fluorescence'

P.R. Hartig, N.J. Bertrand and K. Sauer, (1977), Biochemistry, 16, 4275-82;
'5-Iodoacetamidofluorescein-labelled chloroplast coupling factor 1. Conformational dynamics and labelling-site characterisation'

R.E. Dale, L.A. Chen, and L. Brand, (1977), J. Biol. Chem., 252, 7500-10;
'Rotational relaxation of the 'microviscosity' probe diphenylhexatriene in paraffin oil and egg lecithin vesicles'

W.W. Mantulin and H.J. Pownall, (1977), Photochem. Photobiol., 26, 69-73;
'Fluorescence probes of membrane structure and dynamics'

Ph. Wahl, (1977), Chem. Phys., 22, 245-56;
'Statistical accuracy of rotational correlation times
determined by the photocounting pulse fluorimetry'

F. Hare and C. Lussan, (1977), Biochim. Biophys. Acta,
467, 262-72;
'Variations in microviscosity values induced by different
rotational behaviour of fluorescent probes in some
aliphatic environments'

S. Kawato, K. Kinosita and A. Ikegami, (1977), Biochemistry,
16, 2319-24;
'Dynamic structure of lipid bilayers studied by nanosecond
fluorescence techniques'

Y.S. Borovikov, V.P. Kirillna, N.A. Chernogryadskaya and
A.D. Braun, (1977), Tsitogiya, 19, 382; Chem. Abstr.,
87-3757;
'Study of structural changes in muscle contractile proteins
by polarised ultraviolet fluorescence microscopy. III.
Structural changes in the contractile apparatus of a
muscle fibre during spreading of necrosis'

J.D. Esko, J.R. Gilmore and M. Glaser, (1977), Biochemistry,
16, 1881-90;
'Use of a fluorescent probe to determine the viscosity of
LM cell membranes with altered phospholipid compositions'

L.A. Chen, R.E. Dale, S. Roth and L. Brand, (1977), J. Biol.
Chem., 252, 2163-9;
'Nanosecond time-dependent fluorescence depolarisation of
diphenylhexatriene in dimyristoyllecithin vesicles and
the determination of 'microviscosity''

A. Gafni, R.P. DeToma, R.E. Manrow and L. Brand, (1977),
Biophys. J., 17, 155-68;
'Nanosecond decay studies of a fluorescent probe bound to
apomyoglobin'

R.C. Dorrance and T.F. Hunter, (1977), J. Chem. Soc.,
Faraday I, 73, 1891-99;
'Absorption and emission studies of solubilisation in
micelles. Pt. 4. Studies on cationic micelles with added
electrolyte and on lecithin vesicles : excimer formation
and the Ham effect'

R.E. Pagano, K. Ozato and J-M. Ruysschaert, (1977),
Biochim. Biophys. Acta, 465, 661-3;
'Intracellular distribution of lipophilic fluorescent probes
in mammalian cells'

W.R. Veatch and L. Stryer, (1977), J. Mol. Biol., 117,
1109-13;
'Effects of cholesterol on the rotational mobility of
diphenylhexatriene in liposomes : a nanosecond fluorescence
anisotropy study'

J.R. Lacowicz, D. Hogen and G. Omann, (1977), Biochim.
Biophys. Acta, 471, 401-11;
'Diffusion and partitioning of a pesticide, lindane, into
phosphatidylcholine bilayers. A new fluorescence quenching
method to study chlorinated hydrocarbon-membrane interactions'

K. Kinosita, S. Kawato and A. Ikegami, (1977), Biophys. J.,
20, 289-305;
'A theory of fluorescence polarisation decay in membranes'

M.W. Geiger and N.J. Turro, (1977) Photochem. Photobiol.,
26, 221-4;
'Fluidity and oxygen penetration of lipid vesicles studied
by fluorescence probes'

M. Inbar, N. Larnicol, C. Jasmin, Z. Mishal, Y. Augery,
G. Mather and C. Rosenfeld, (1977), Eur. J. Cancer,
13, 1231-6;
'A method for the quantitative detection of human acute
lymphatic leukemia'

M.R. Eftink and C.A. Ghiron, (1977), Biochemistry, 16,
5546-51;
'Exposure of tryptophanyl residues and protein dynamics'

M.P. Heyn, R.J. Cherry and U. Muller, (1977), J. Mol. Biol.,
117, 607-620;
'Transient and linear dichroism studies on bacteriorhodopsin : determination of the orientation of the 568 nm
all-trans retinal chromophore'

S.M. Johnson and C. Nicolau, (1977), Biochem. Biophys. Res.
Comm., 76, 869-874;
'The distribution of 1,6-diphenyl hexatriene fluorescence
in normal human lymphocytes'

C.A. King and T.M. Preston, (1977), FEBS Letters, 73, 59-63;
'Fluoresceinated cationized ferritin as a membrane probe for
anionic sites'

R.F. Chen, (1977), Arch. Biochem. Biophys., 179, 672-81;
'Fluorescence of free and protein-bound auramine O'

E.S. Tecoma, L.A. Sklar, R.D. Simoni and B.S. Hudson, (1977)
Biochemistry, 16, 829-35;
'Conjugated polyene fatty acids as fluorescent probes :
biosynthetic incorporation of parinaric acid by E. Coli
and studies of phase transitions'

J. Breton, J. Viret and F. Leterrier, (1977), Arch. Biochem.
Biophys., 179, 625-33;
'Calcium and chlorpromazine interactions in rat synaptic
plasma membranes. A spin label and fluorescence probe study'

L.A. Sklar, B.S. Hudson, M. Petersen and J. Diamond, (1977),
Biochemistry, 16, 813-19;
'Conjugated polyene fatty acids as fluorescent probes:
spectroscopic characterisation'

L.A. Sklar, B.S. Hudson and R.D. Simoni, (1977), Biochemistry,
16, 819-28;
'Conjugated polyene fatty acids as fluorescent probes:
synthetic phospholipid membrane studies'

Y. Kanoaka, (1977), Angew. Chem., 89, 142-52;
'Organic fluorescence reagents for the investigation of
enzymes and proteins'

F. Podo and J.K. Blasie, (1977), Proc. Natl. Acad. Sci.
U.S.A., 74, 1032-6;
'Nuclear magnetic resonance studies of lecithin bimolecular
leaflets with incorporated fluorescent probes'

S.L. Helgerson, (1977), Diss. Abstr. Int. B., 37, 4836-7;
'The E.Coli cell envelope : a fluorescence probe study of the
in vivo energy dependent structure'

S. Takeuchi and A. Maeda, (1977), J. Biochem., 81, 971-6;
'Use of fluorescein mercuric acetate as a probe in studies
of thiol-containing proteins'

D. Nieva-Gomez and R.B. Gennis, (1977), Proc. Natl. Acad.
Sci. U.S.A., 74, 1811-15;
'Affinity of intact E. Coli for hydrophobic membrane probes'

M.W. Geiger, (1977), Diss. Abstr. Int. B., 37, 5083-4;
'Part 1. The spectroscopy and photochemistry of brilliant
green leucocyanide.   Part 2. Fluorescence probes in model
membrane systems'

D.H. Haynes and P. Simkowitz, (1977), J. Membr. Biol., 33,
63-108;
'1-anilino-8-naphthalenesulphonate : a fluorescent probe of
ion and ionophore transport kinetics and trans-membrane
asymmetry'

M. Lewin, G. Saccomani, R. Schackmann and G. Sachs, (1977),
J. Membr. Biol., 32, 301-18;
'Use of 1-anilino-8-naphthalenesulphonate as a probe of
gastric vesicle transport'

A.I. Deev, E.A. Yarova and G.E. Dobretsov, (1977),
Farmakol. Toksikol (Moscow), 40, 351-5; Chem. Abstr., 87-
62340;
'Fluorescent probe study of the binding of phenothiazines
to membranes'

N.S. Ranadive and D.H. Ruben, (1977), Immunochemistry, 14,
165-70;
'Mast cell-lysomal cationic protein interaction: use of
1-anilinonaphthalene-8-sulphonate as a fluorescent probe'

E. Koller, T. Vukovich, F. Koller and W. Doleschel, (1977),
Wien. Klin. Wochenschr., 89, 379-82; Chem. Abstr., 87-80690;
'Lipoprotein analysis using the fluorescent probe 8-anilino-
1-naphthalenesulphonic acid'

Y.J. Lee, A.C. Notides, Y-G. Tsay and A.S. Kende (1977),
Biochemistry, 16, 2896-901;
'Coumestrol, NBD-norhexestrol, and dansyl-norhexestrol,
fluorescent probes of estrogen-binding proteins'

A.G. Sabel'nikov, T.F. Moiseeva, A.V. Audeeva and B.N.
Il'yashenko, (1977), Biofizika, 22, 640-5; Chem. Abstr.,
22, 640-5;
'Study of the interaction between fluorescent probe and
E. Coli surface structures'

G.F.W. Searle, J. Barber and J.D. Mills, (1977), Biochim.
Biophys. Acta, 461, 413-25;
'9-amino-acridine as a probe of the electrical double layer
associated with the chloplast thylakoid membranes'

R. Kraayenhof and J.C. Arents, (1977), Electr. Phenom. Biol.
Membr. Level, Proc. Int. Meet. Soc. Chim. Phys., 29th,
439-505; Chem. Abstr., 87-130663;
'Fluorescent probes for the chloroplast energised state -
energy-linked change of membrane-surface charge'

S.L. Helgerson and W.A. Cramer, (1977), Biochemistry, 16,
4109-17;
'Changes in E.Coli cell envelope structure and the sites of
fluorescence probe binding caused by carbonyl cyanide
p-trifluoromethoxyphenylhydrazone'

G.E. Dobretsov, V.A. Petrov, V.E. Mishizhev, G.I. Klebanov
and Y.A. Vladimorov, (1977), Stud. Biophys., 65, 91-8;
Chem. Abstr. 87-148166;
'4-dimethylaminochalcone and 3-methoxybenzanthrone as
fluorescent probes to study biomembranes 1. Spectral
characteristics'

B.M.J. Kellner, (1977), Diss. Abstr. Int. B, 38, 1232;
'Pure and mixed monomolecular films of fluorescent probes
of cell membranes'

R.D. O'Brien, T.J. Herbert and B.D. Hilton, (1977),
Pestic. Biochem. Physiol., 7, 416-25; Chem. Abstr., 87-
195045;
'The effects of DDT upon flat and vesicular bilayers studied
by a fluorescent probe'

D. Cadenhead, B.M.J. Kellner, K. Jacobson and D. Papahadjopoulos, (1977), Biochemistry, 16, 5386-92;
'Fluorescent probes in model membranes 1. Anthroyl fatty
acid derivatives in monolayers and liposomes of dipalmitoylphosphatidylcholine'

J.A. Monti, S.T. Christian, W.A. Shaw and W.H. Finley, (1977),
Life Sci., 21, 345-55;
'Synthesis and properties of a fluorescent derivative of
phosphatidylcholine'

L.A. Sklar, B.S. Hudson and R.D. Simoni, (1977), Biochemistry,
16, 5100-8;
'Conjugated polyene fatty acids as fluorescent probes :
binding to bovine serum albumin'

P.R. Adams and A. Feltz, (1977), Nature, 269, 609-11;
'Interaction of a fluorescent probe with acetylcholine-
activated synaptic membrane'

S. Schuldiner and R.H. Kaback, (1977), Biochim. Biophys.
Acta, 472, 399-418;
'Fluorescent galactosides as probes for the lac carrier
protein'

A. Bruni, B. Tosi and G. Dall'Olio, (1977), Histochem. J.,
9, 703-9;
'Fluorescamine : a fluorescent probe for amino groups in
histochemical studies on plant cells and the effects of
mercury fixation'

K. Zierler, (1977), Biophys. Struct. Mech., 3, 275-89;
'An error in interpretation of double-reciprocal plots and
Scatchard plots in studies of binding of fluorescent probes
to proteins, and alternative proposals for determining
binding parameters'

U. Prasad, G.S. Singwal and P. Mohanty, (1977), Biophy.
Struct. Mech., 3, 259-74;
'Effects of protons and cations on chloroplast membranes as
visualised by the bound ANS fluorescence'

E.V. Nikushkin, G.I. Klebanov and Y.A. Vladimirov, (1977),
Biofizika, 22, 1049-55; Chem. Abstr., 88-59891;
'Phase transition studies of model and biological membranes.
1. Use of hydrophobic fluorescent probes for phase transition
studies in liposomes'

K. Nowak, M. Kuczek, A. Slominska and T. Wilusz, (1977),
Acta Biochim. Pol., 24, 275-80; Chem. Abstr., 88-166012;
'Amino acid sequence of the basic trypsin inhibitor from
bovine splenic capsule. A new fluorescent reagent from
protein sequence analysis'

P.M. Keller, S. Person and W. Snipes, (1977), J. Cell. Sci.,
28, 167-77;
'A fluorescent enhancement assay of cell fusion'

D. Atlas and A. Levitzki, (1977), Proc. Natl. Acad. Sci.
U.S.A., 74, 5290-4;
'Probing of beta-adrenergic receptors by novel fluorescent
beta-adrenergic blockers'

D.E. Chandler and J.A. Williams, (1977), Nature, 268, 659-60;
'Fluorescent probe detects redistribution of cell calcium
during stimulus-secretion coupling'

D. Njus, S.J. Ferguson, M. Sorgato and G.K. Radda, (1977),
BBA Libr., 14, (Struct. Funct. Energy-Transduing Membr.)
237-50; Chem. Abstr., 89-55651;
'The ANS response, eight years later'

T.E. Thompson, B.R. Lentz and Y. Barenholz, (1977), FEBS-
Symp. 42 (Biochem. Membr. Transport), 47-71; Chem. Abstr.
87-97445;
'A colorimetric and fluorescent probe study of phase
transitions in phosphatidylcholine liposomes'

D.E. Chandler and J.A. Williams, (1978), J. Cell. Biol.,
76, 386-99;
'Intracellular divalent cation release in pancreatic acinar
cells during stimulus-secretion coupling. II Subcellular
localisation of the fluorescent probe chlorotetracycline'

C.D. Tran, P.L. Klahn, A.Romero and J.H. Fendler, (1978),
J. Amer. Chem. Soc., 100, 1622-4;
'Characterisation of surfactant vesicles as potential membrane
models. Effect of electrolytes, substrates and fluorescent
probes'

W.C. Horne and E.R. Simons, (1978), Blood, 51, 745-53;
'Probes of transmembrane potentials in platelets : changes
in cyanine dye fluorescence in response to aggregation
stimuli'

L.G. Korkina, G.E. Dobretsov, G. Walzel, E.M. Kogan and
Y.A. Vladimirov, (1978), Dokl. Akad. Nauk. SSSR, 238, 999-
1002; Chem. Abstr., 88-134619;
'Identification of T- and B-lymphocytes using a membrane
fluorescent 3-methoxybenzanthrone probe'

N.S. Kosower, E.M. Kosower, S. Lustig and D.H. Pluznik,
(1978), Biochim. Biophys. Acta, 507, 128-36;
'F2OC, a new fluorescent membrane probe, moves more slowly
in malignant and mitogen-transformed cell membranes than
in normal cell membranes'

R.P. Liburdy, (1977), J. Phys. Chem., 82, 870-4;
'N-(3-pyrene) succinimidothioethanol. A fluorescent probe
sensitive to pH and redox potential'

C.A.M. Carvalho, (1978), J. Neurochem., 30, 1149-55;
'Chlorotetracycline as an indicator of the interaction of
calcium with brain membrane fractions'

B.R. Masters, (1977, pub. 1978), Biomol. Struct. Funct.
(Symp.), 123-7 ; Chem. Abstr., 89-71250;
'Fluorescent probe study of antidiuretic hormone-induced
changes in membrane fluidity and water permeability'

W.W. Webb, (1978), Front. Biol. Energ., 2, 1333-9; Chem.
Abstr., 91-85390;
'Features and function of lateral motion on cell membrane
revealed by fluorescence dynamics'

Y.S. Borovikov, (1978), Biofiz. Biokhim. Metody. Issled.
Myshechnykh Belkov, 223-36; Chem. Abstr., 91-71020;
'Use of polarising UV microscopy for studying contractile
proteins of muscle fibre'

Ph. Wahl, K. Tawada amd J.C. Auchet, (1978), Eur. J.
Biochem., 88, 421-4;
'Study of tropomysin labelled with a fluorescent probe by
pulse fluorimetry in polarised light. Interaction of that
protein with troponin and actin'

K. Tawada, Ph. Wahl and J.C. Auchet, (1978), Eur. J.
Biochem., 88, 411-19;
'Study of actin and its interactions with heavy meromyosin
and the regulatory proteins by the pulse fluorimetry in
polarised light of a fluorescent probe attached to an actin
cysteine'

C.E. Martin and G.A. Thompson, (1978), Biochemistry, 17,
3581-6;
'Use of fluorescence polarisation to monitor intercellular
membrane changes during tenperature acclimation. Correlation with lipid compositional and ultrastructural changes'

K.R. Thulborn and W.H. Sawyer, (1978), Biochim. Biophys.
Acta, 511, 125-40;
'Properties and locations of a set of fluorescent probes
sensitive to the fluidity gradient of the lipid bilayer'

B.A. Smith and H.M. McConnell, (1978), Proc. Natl. Acad.
Sci. U.S.A., 75, 2759-63;
'Determination of molecular motion in membranes using
periodic pattern photobleaching'

C.D. Stubbs, W.M. Tsang, J. Belin, A.D. Smith and S.M.
Johnson, (1978), Biochem. Soc. Trans., 6, 289-90;
'Effect of rat lymphocytes of fatty acids and concanavalin A

on the fluorescence polarisation of 1,6-diphenyl hexatriene'

D.F. Hahherty, V.K. Kalra, G. Popjak, E.E. Reynolds and
F. Chiappelli, (1978), Arch. Biochem. Biophys., 189, 51-62;
'Fluorescence polarisation measurements on normal and
mutant human skin fibroblasts'

J.R. Lakowicz and F.G. Prendergast, (1978), Science, 200,
1399-1401;
'Quantitation of hindered rotations of diphenylhexatriene
in lipid bilayers by differential polarised phase
fluorometry'

G.D. Correll, R.N. Cheser, F. Nome and J.H. Fendler, (1978),
J. Amer. Chem. Soc., 100, 1254-62;
'Fluorescence probes in reversed micells. Luminescence
intensities, lifetimes, quenching, energy transfer, and
depolarisation of pyrene derivatives in cyclohexane in the
presence of dodecylammonium propionate aggregates'

E.I. Dudich, R.S. Nezlin, and F. Franek, (1978), FEBS
Letters, 89, 89-92;
'Fluorescence polarisation analysis of various immunoglobulins. Dependence of rotational relaxation time
on protein concentration and on ability to precipitate
with antigen'

H. Shimonaka, H. Fukushima, K. Kawai, S. Nagao, Y. Okano
and Y. Nozawa, (1978), Experientia, 34, 586-7;
'Altered microviscosity of in vivo lipid-manipulated
membranes in Tetrahymena Pyriformis : a fluorescence
study'

A. Obrenovitch, C. Sene, M.T. Negre and M. Monsigny,
(1978), FEBS Letters, 88, 187-91;
'Fluorescence polarisation of 1,6-diphenyl-1,3,5-hexa-
triene embedded in membranes of mouse leukemic L 1210
cells during the cell cycle'

E.A. Burshtein, T.L. Bushueva and E.A. Permyakov, (1978),
Zh. Prikl. Spektrosk., 28, 653-7;
'Study of the equilibrium structural mobility of macromolecules according to the luminescence characteristics
of protein chromophores'

J.H. Easter, (1978), Diss. Abstr. Int. B, 39, 2630;
'A nanosecond glimpse at fluorescent probes bound to model
membranes'

S.L. Rosenthal, A.H. Paraola, E.R. Blout and R.L. Davidson, (1978), Exp. Cell. Res., 112, 419-29; 'Membrane alterations associated with 'transformation' by BUdR in BUdR-dependent cells. Fluorescence polarisation studies with a lipid probe'

J.R. Lepock, J.E. Thompson, J. Kruuv and D.F.H. Wallach, (1978), Biochem. Biophys. Res. Comm., 85, 344-50; 'Photoinduced crosslinking of membrane proteins by fluorescein isothiocyanate'

M. Shinitzky and Y. Barenholz, (1978), Biochim. Biophys. Acta, 515, 367-94; 'Fluidity parameters of lipid regions determined by fluorescence polarisation'

G.W. Robinson, R.J. Robbins, G.R. Fleming, J.M. Morris, A.E.W. Knight and R.J.S. Morrison, (1978), J. Amer. Chem. Soc., 100, 7145-50; 'Picosecond studies of the fluorescent probe molecule 8-anilino-1-naphalenesulphonic acid'

D.D. Thomas, (1978), Biophys. J., 24, 439-62; 'Large scale rotational motions of proteins detected by electron paramagnetic resonance and fluorescence'

G.R. Burleson, C.F. Kulpa, H.E. Edwards and J.K. Thomas, (1978), Exp. Cell. Res., 116, 291-300; 'Fluorescent probe studies of normal, persistently infected, Rous Sarcoma virus-transformed and trypsinised rat cells'

Y.S. Borovikov, V.P. Kirillina and N.A. Chernogryadskaya, (1978), Tsitologiya, 20, 1161-6; Chem. Abstr., 90-20104; 'Study of structural changes in muscle fibre contractile proteins by polarised ultraviolet fluorescent microscopy. IV. Some features of F-actin conformational changes in muscle fibre relaxation'

F. Hare and C. Lussan, (1978), FEBS Letters, 94, 231-5; 'Mean viscosities in microscopic systems and membrane bilayers. A semi-empirical general basis applicable to different kinds of extrinsic probes'

A.C. Roche, R. Maget-Dana, A. Obrenovitch, H. Knut, C. Nicolau and M. Monsigny, (1978), FEBS Letters, 93, 91-6; 'Interaction between vesicles containing gangliosides and limulin (Limulus Polyphemus Lectin). Fluorescence polarisation of 1,6-diphenyl-1,3,5-hexatriene'

S. Kawato, K. Kinosita and A. Ikegami, (1978), Biochemistry,
17, 5026-31;
'Effect of cholesterol on the molecular motion in the hydrocarbon region of lecithin bilayers studied by nanosecond
fluorescence techniques'

R.P. DeToma and L. Brand, (1978), Biophys. J., 24, 197-212;
'Nanosecond relaxation processes in liposomes'

G. Thomas, J.L. Fourrey and A. Faure, (1978), Biochemistry,
17, 4500-8;
'Reduced 8-13 link, a viscosity-dependent fluorescent probe
of transfer RNA tertiary structure'

P.J. Stein and R.W. Henkens, (1978), J. Biol. Chem., 253,
8016-18;
'Detection of intermediates in protein folding of carbonic
anhydrase with fluorescence emission and polarisation'

S.S. Gupte, (1978), Diss. Abstr. Int. B., 39, 1099-1100;
'Fluidity and chromophore interactions in purple membrane :
electron spin resonance, fluorescence and circular dichroism
study'

Y.Y. Leshem and M. Inbar, (1978), J. Exp. Bot., 29, 671-5;
'Resistance to gibberellin-induced changes of lipid
fluidity in wheat embryo mitochondrial membranes as
assessed by the fluorescent probe 1,6-diphenyl-1,3,5-
hexatriene'

P. Glatz, (1978), Anal. Biochem., 87, 187-94;
'Limited rotational diffusion of DPH in human erythrocyte
membranes'

G. Berke, R. Tzar and M. Inbar, (1978), J. Immunol., 120,
1378-1384;
'Changes in fluorescence polarisation of a membrane probe
during lymphocyte-target cell interaction'

P.R. Dragsten, (1978), Diss. Abstr. Int. B, 38, 4026;
'Mechanism of voltage-induced fluorescence changes of
the membrane probe merocyanine 540. A fluorescence
polarisation study'

J.M. Freyssinet, B.A. Lewis, J.J. Holbrook and J.D. Shore,
(1978), Biochem. J., 169, 403-10;
'Protein-protein interactions in blood clotting. The use
of polarisation of fluorescence to measure the dissociation
of plasma factor XIIIa'

G. Laustriat and D. Gerard, (1978), J. Phys. Chem., 82,
746-9;
'Influence of dynamic quenching on the thermal dependence
of fluorescence in solution. Study of indole and phenol
in water and dioxane'

J. Teissie, J.F. Tocanne and A. Baudras, (1978), Eur. J.
Biochem., 83, 77-85;
'A fluorescence approach of the determination of translational diffusion coefficients of lipids in phospholipid
monolayer at the air-water interface'

R.D. Philo and A.A. Eddy, (1978), Biochem. J., 174, 801-10;
'The membrane potential of mouse ascites-tumor cells
studied with the fluorescent probe 3,3'-dipropyloxa-
dicarbocyanine. Amplitide of the depolarisation caused
by amino acids'

Y.S. Borovikov, V.P. Kirillina and N.A. Chernogryadskaya,
(1978), Tsitologiya, 20, 1384-9; Chem. Abstr., 90-84561;
'Polarised ultraviolet fluorescence study of structural
changes in muscle contractile proteins. V. Possible nature
of heavy meromyosin conformational changes during fibre
relaxation'

D.J. De La Motte, (1978), Exp. Eye Res., 27, 585-94;
'Removal of horseradish peroxidase and fluorescein-
labelled dextran from CSF spaces of rabbit optic nerve.
A light and electron microscope study'

H. Maeda, (1978), Kagaku to Seibutsu, 16, 729-35; Chem.
Abstr., 90-83012;
'Fluorescence polarisation and its applications'

H. Maeda, (1978), Clin. Chem., 24, 2139-44;
'Assay of an antitumor protein, neocarzinostatin, and its
antibody by fluorescence polarisation'

F.J.L. Rodier, (1978), Biochimie, 60, 609-17;
'Study and comparison of the polarity of the active site
of three serine proteases using a fluorescent probe'

J.A. Monti, S.T. Christian and W.A. Shaw, (1978), J. Lipid
Res., 19, 222-8;
'Synthesis and properties of a highly fluorescent deriv-
ative of phosphatidylethanolamine'

T.G. Easton, J.E. Valinsky and E. Reich, (1978), Cell
(Cambridge, Mass.), 13, 475-86;
'Merocyanine 540 as a fluorescent probe of membranes :
staining of electrically excitable cells'

J.E. Valinsky, T.G. Easton and E. Reich, (1978), Cell
(Cambridge, Mass.), 13, 487-99;
'Merocyanine 540 as a fluorescent probe of membranes :
Selective staining of leukemic and immature hemopoietic
cells'

V.T. Maddaiah, M. Kumbar and P.J. Collipp, (1977), Biomol.
Struct. Funct., (Symp.), 129-36, Academic Press, New York;
Chem. Abstr., 89-38740;
'Interaction of N-phenyl-1-naphthyl amine and 1-anilino-
8-naphthalene sulphonate with glucose 6-phosphatase of
hepatic microsomes'

G. Dreyfuss, K. Schwartz, E.R. Blout, J.R. Barrio, F-T. Liu
and N.J. Leonard, (1978), Proc. Natl. Acad. Sci. U.S.A.,
75, 1199-203;
'Fluorescent photoaffinity labelling : adenosine 3',5'-
cyclic monophosphate receptor sites'

C. Seny, D. Genest, A. Obrenovitch, P. Wahl and M. Monsigny,
(1978), FEBS Letters, 88, 181-6;
'Pulse fluorimetry of 1,6-diphenyl-1,3,5-hexatriene
incorporated in membranes of mouse leukemic L 1210 cells'

F. Schroeder and J.F. Holland, (1978), Biomol. Struct.
Funct., (Symp.), 137-45, Academic Press, New York; Chem.
Abstr., 89-39883;
'Fluorescent probes and the structure of mammalian membranes'

S.L. Betcher-Lange and S.S. Lehrer, (1978), J. Biol. Chem.,
253, 3757-60;
'Pyrene excimer fluorescence in rabbit skeletal $\alpha.\alpha.$ tropomyosin labelled with N-(1-pyrene) maleimide. A probe of
sulphydryl proximity and local chain separation'

I.A. Bailey, C.J. Garratt and S.M. Wallace, (1978),
Biochem. Soc. Trans., 6, 302-4;

'An effect of fluorescent probes and of insulin on the structure of adipocyte membranes'

J. Rogers, A.G. Lee and D.C.Wilton, (1978), Biochem. Soc. Trans., 6, 281-4;
'A proposed structure for cholesterol-containing bilayers based on studies with fluorescent sterols'

S. Yonei and M. Kato, (1978), Radiat. Res., 75, 31-45;
'X-ray induced structural changes in erythrocyte membranes studied by use of fluorescent probes'

J.G. Collard, A. De Wildt and M. Inbar, (1978), FEBS Letters, 90, 24-8;
'Translocation of a fluorescent lipid probe between contacting cells. Evidence for membrane lipid interactions'

F.J. Farris, G. Weber, C.C. Chiang and I.C. Paul, (1978), J. Amer. Chem. Soc., 100, 4469-74;
'Preparation, crystalline structure, and spectral properties of the fluorescent probe 4,4'-bis-1-phenylamino-8-naphthalene sulphonate'

S. Cheng, H.M. McQueen and D. Levy, (1978), Arch. Biochem. Biophys., 189, 336-43;
'The interaction of calcium and procaine with hepatocyte and hepatoma tissue culture cell plasma membranes studied by fluorescence spectroscopy'

U. Schummer, H.G. Schiefer and U. Gerhardt, (1978), Hoppe-Seyler's Z. Physiol. Chem., 359, 1023-5; Chem. Abstr., 89-159876;
'Mycoplasma membrane potential determined by a fluorescent probe'

Y. Katsumata, O. Suzuki and M. Oya, (1978), FEBS Letters, 93, 58-60;
'Changes in the mean distance between tryptophan and fluorescence probe in the labelled sarcoplasmic reticulum membranes induced by detergents'

E. Lazarides and B.L. Granger, (1978), Proc. Natl. Acad. Sci. U.S.A., 75, 3683-7;
'Fluorescent localisation of membrane sites in glyerinated chicken skeletal muscle fibres and the relationship of these sites to the protein composition of the Z disc'

214 BIBLIOGRAPHY 1978

A.E. Zakaryan, E.S. Sekoyan and A.R. Egiazaryan, (1978),
Biol. Zh. Arm., 31, 527-32; Chem. Abstr., 90-2038;
'Study of the interaction of the fluorescent probe 1-anilino-
naphthalene-8-sulphonate with plasma membrane preparations'

P.R. Dragsten and W.W. Webb, (1978), Biochemistry, 17, 5228-40;
'Mechanism of the membrane potential sensitivity of the
fluorescent membrane probe merocyanine 540'

G.E. Dobretsov, V.M. Dmitriev, L.B. Pirogova, V.A. Petrov
and Y.A. Vladimirov, (1978), Stud. Biophys., 71, 189-96;
Chem. Abstr., 90-17946;
'4-Dimethylaminochalcone and 3-methoxybenzanthrone as
fluorescent probes to study biomembranes. III. Relationship
between state of hydration shell of membrane and phase state
of phospholipids'

G.E. Dobretsov, V.A. Petrov and Y.A. Vladimirov, (1978),
Stud. Biophys., 71, 181-7; Chem. Abstr., 90-18553;
'4-Dimethylaminochalcone and 3-methoxybenzanthrone as
fluorescent probes to study biomembranes. II. Sensitivity of
4-dimethylaminochalcone to water molecules in the surface
layer of a membrane'

T. Horie, Y. Sugiyama, S. Awazu and M. Hanano, (1978),
J. Pharmacobio-Dyn., 1, 203-12; Chem. Abstr., 90-48307;
'The correlation between drug binding to the human erythrocyte
membrane and its hemolytic activity determined by fluorescent
probe techniques'

K. Zierler and E. Rogus, (1978), Biochim. Biophys. Acta,
514, 37-53;
'Fluorescence of 1,8-anilinonaphthalene sulphonic acid bound
to proteins and to lipids of sarcolemma'

E.A. Haigh, K.R. Thulborn, L.W. Nicol and W.H. Sawyer, (1978),
Aust. J. Biol. Sci., 31, 447-57;
'Uptake of N-(9-anthroyloxy) fatty acid fluorescent probes
into lipid bilayers'

B.M.J. Kellner and D.A. Cavenhead, (1978), Biochim. Biophys.
Acta, 513, 301-9;
'Fluorescent probes in model membranes. II. Monolayer studies
of 2,2'-(vinylenedi-p-phenylene) bisbenzoxazole, d-3-amino-
desoxyequilenin and N-octadecylnaphthyl-2-amino-6-sulphonic
acid with host-lipid tetradecanoic acid'

J.D. Mills and G. Hind, (1978), Photochem. Photobiol., 28, 67-73;
'Use of the fluorescent lanthanide terbium $^{3+}$ as a probe for cation-binding sites associated with isolated chloroplast thylakoid membranes'

L. Spero, (1978), Can. J. Physiol. Pharmacol., 56, 915-20;
'A study of carbachol-atropine interaction on intestinal smooth muscle vesicles using a fluorescent probe'

C. Holzapfel, (1978), Ber. Kernforschungsanlage Juelich, (Juel-1542); Chem. Abstr., 90-164206;
'Study of biological membranes using ANS as a fluorescence sample'

D. Georgescauld and H. Duclohier, (1978), Biochem. Biophys. Res. Comm., 85, 1186-91;
'Transient fluorescence signals from pyrene-labelled pike nerves during action potential. Possible implications for membrane fluidity changes'

K. Kano and J.H. Fendler, (1978), Biochim. Biophys. Acta, 509, 289-299;
'Pyranine as a sensitive pH probe for liposome interiors and surface. pH gradients across phospholipid vesicles'

J.R. Lakowicz and F.G. Prendergast, (1979), J. Biol. Chem., 254, 1771-4;
'Nanosecond relaxation in membranes observed by fluorescence lifetime-resolved emission spectra'

B.R. Lentz, B.M. Moore and D.A. Barrow, (1979), Biophys. J., 25, 489-94;
'Light scattering effects in the measurement of membrane microviscosity with diphenylhexatriene'

G. Paillotin, C.E. Swenberg, J. Breton and N.E. Gaecintov, (1979), Biophys. J., 25, 513-33;
'Analysis of picosecond laser-induced fluorescence phenomena in photosynthetic membranes utilising a master equation approach'

R.P. Van Hoeven, W.J. Van Blitterswijk and P. Emmelot, (1979), Biochim. Biophys. Acta, 551, 44-54;
'Fluorescence polarisation measurements on normal and tumor cells and their corresponding plasma membranes'

F. Schroeder, E.H. Goh and M. Heimberg, (1979), FEBS Letters, 97, 233-6;
'Investigation of the surface structure of the very low density lipoprotein using fluorescent probes'

K. Zierler and E.M. Rogus, (1979), Biochim. Biophys. Acta, 551, 389-405;
'Temperature effects on 1,8-anilinonaphthalene sulphonic acid fluorescence with sarcolemma vesicles'

T. Ohyashiki and T. Mohri, (1979), J. Biochem. (Tokyo), 85, 857-63; Chem. Abstr., 90-148014;
'1-anilino-8-naphthalene sulphonate as a probe of conformational changes of rabbit intestinal brush border membranes on addition of salts and sugars'

G.F.W. Searle and J. Barber, (1979), Biochim. Biophys. Acta, 545, 508-18;
'The interaction of an amphipathic fluorescence probe, 2-p-toliudinonaphthalene-6-sulphonate, with isolated chloroplasts'

W.R. Laws, G.H. Posner and L. Brand, (1979), Arch. Biochem. Biophys., 193, 88-100;
'A covalent   fluorescence probe based on excited-state proton transfer'

J. Rogers, A.G. Lee and D.C. Wilton, (1979), Biochim. Biophys. Acta, 552, 23-37;
'The organisation of cholesterol and ergosterol in lipid bilayers based on studies using non-perturbing fluorescent sterol probes'

F. Schroeder, E.H. Goh and M. Heimberg, (1979), J. Biol. Chem., 254, 2456-63;
'Regulation of the surface physical properties of the very low density lipoprotein'

F. Schroeder and E.H. Goh, (1979), J. Biol. Chem., 254, 2464-70;
'Regulation of very low density lipoprotein interior core lipid physicochemical properties'

M. Schimerlik, U. Quast and M.A. Raftery, (1979), Biochemistry; 18, 1884-90;
'Ligand-induced changes in membrane-bound acetylcholine receptor observed by ethidium fluorescence. 1. Equilibrium studies'

J. Luisetti, H. Moehwald and H.J. Galla, (1979), <u>Biochim</u>.
<u>Biophys</u>. <u>Acta</u>, <u>552</u>, 519-30;
'Monitoring the location profile of fluorophors in phosphatidylcholine bilayers by the use of paramagnetic quenching'

R.M. Smillie, (1979), <u>Aust</u>. <u>J</u>. <u>Plant Physiol</u>., <u>6</u>, 121-33;
'Coloured components of chloroplast membranes as intrinsic
membrane probes for monitoring the development of heat
injury in intact tissues'

C, Montecucco, T. Pozzan and T. Rink, (1979), <u>Biochim</u>.
<u>Biophys</u>. <u>Acta</u>, <u>552</u>, 552-7;
'Dicarbocyanine fluorescent probes of membrane potential
block lymphocyte capping : deplete cellular ATP and inhibit
respiration of isolated mitochondria'

S. Ozeki and K. Tejima, (1979), <u>Chem</u>. <u>Pharm</u>. <u>Bull</u>., <u>27</u>,
638-46;
'Drug interactions. V. Binding of basic compounds to bovine
serum albumin by fluorescent probe technique'

J. Yguerabide and M.C. Foster, (1979), <u>J</u>. <u>Membr</u>. <u>Biol</u>., <u>45</u>,
109-23;
'Theory of lipid bilayer phase transitions as detected by
fluorescent probes'

J.R. Escabi-Perez, (1979), <u>Diss</u>. <u>Abstr</u>. <u>Int</u>. <u>B</u>, <u>39</u>, 4891;
'Fluorescence probes for the investigation of membrane
mimetic systems'

J. Slavik, (1979), <u>Biol</u>. <u>Listy</u>, <u>44</u>, 10-26; <u>Chem</u>. <u>Abstr</u>.,
<u>91</u>-16008;
'Use of fluorescent probes in the study of biological
membranes'

Y.G. Molotkovskii, P. Dmitriev, L.F. Nikulina and
L.D. Bergelson, (1979), <u>Bioorg</u>. <u>Khim</u>., <u>5</u>, 588-94; <u>Chem</u>.
<u>Abstr</u>., <u>91</u>-52111;
'Synthesis of new fluorescent-labelled phosphatidylcholines'

G.A. Gibson and L.M. Loew, (1979), <u>Biochem</u>. <u>Biophys</u>. <u>Res</u>.
<u>Commun</u>., <u>88</u>, 135-40;
'Phospholipid vesicle fusion monitored by fluorescence energy
transfer'

J. Kovar, (1979), <u>Chem</u>. <u>Listy</u>, <u>73</u>, 614-38;
'Use of fluorescence probes in the study of proteins'

R.D. Brown and K.S. Matthews, (1979), <u>J. Biol. Chem.</u>,
<u>254</u>, 5135-43;
'Spectral studies on LAC repressor modified with N-substit-
uted maleimide probes'

C. Graue and M. Klingenberg, (1979), <u>Biochim. Biophys. Acta</u>,
<u>546</u>, 539-50;
'Studies of the ADP/ATP carrier of mitochondria with
fluorescent ADP analogue formycin diphosphate'

M.D. Baratt and A.R. Badley, (1979), <u>Biochem. Dis.</u>, <u>7</u>
(Biochem. Atheroscler.), 75-106; <u>Chem. Abstr.</u>, <u>91</u>-71022;
Lipid-protein interactions with particular reference to
fluorescence and electron spin resonance spectroscopy'

B. De Foresta, T. Nguyen Le, C. Nicot and A. Alfsen, (1979)
<u>Biochimie</u>, <u>61</u>, 522-33;
'Study of fluorescent tryptophyl residues and extrinsic
probes for the characterisation of molecular domains
of Folch-pi apoprotein'

A. Azzi, (1979), <u>Methods Enzymol.</u>, <u>56</u> (Biomembranes, part G),
486-501;
'Use of fluorescence to study energy-linked processes'

A.G. Sabel'nikov and B.N. Il'yashenko, (1979), <u>Byull. Eksp.</u>
<u>Biol. Med.</u>, <u>88</u>, 65-8; <u>Chem. Abstr.</u>, <u>91</u>-136501;
'Dimethylaminochalcon - a probe for structural changes in
E.Coli cell envelopes induced by calcium ion and tris-
buffer treatment'

G.M. Halliday, R.C. Nairn, M.A. Pallett, J.M. Rolland and
H. Ward, (1979), <u>J. Immunol. Methods</u>, <u>28</u>, 381-90;
'Detection of early lymphocyte activation by the fluorescent
cell membrane probe N-phenyl-1-naphthylamine'

J.G. Hoggett, (1979), <u>Amino-acids, pept., proteins</u>, <u>10</u>,
291-309; <u>Chem. Abstr.</u>, <u>91</u>-152857;
'Structural investigations of peptides and proteins. III.
Conformation and interaction of peptides and proteins in
solution. 9. Fluorescence'

A.G. Gabibov, S.N. Kochetkov, L.P. Sashchenko and E.S. Severin,
(1979), <u>Biochim. Biophys. Acta</u>, <u>569</u>, 145-52;
'Determination of binding parameters of cyclic AMP and its
analogues to cyclic AMP-dependent protein kinase by the
fluorescent probe method'

G.P. Sachdev, J.M. Zodrow and R. Carubelli, (1979), Biochim. Biophys. Acta, 580, 85-90;
'Hydrophobic interaction of fluorescent probes with fetuin, ovine submaxillary mucin, and canine tracheal mucins'

L.A. Sklar, G.P. Miljanich and E.A. Dratz, (1979), J. Biol. Chem., 254, 9592-7;
'A comparison of the effects of calcium on the structure of bovine retinal rod outer segment membranes, phospholipids, and bovine phosphatidylserine'

I. Nathan, G. Fleischer, A. Livne, A. Dvilansky and A.H. Parola, (1979), J. Biol. Chem., 254, 9822-8;
'Membrane microenvironmental changes during activation of human blood platelets by thrombin. A study with a fluorescent probe'

A.H. Parola, P.W. Robbins and E.R. Blout, (1979), Exp. Cell Res., 118, 205-14;
'Membrane dynamic alterations associated with viral transformation and reversion. Decay of fluorescence emission and anisotropy studies with 3T3 cells'

D.L. Barbeau, A. Jonas, T-L. Teng and A.M. Scanu, (1979), Biochemistry, 18, 362-9;
'Asymmetry of apolipoprotein A-1 in solution as assessed from ultracentrifugal, viscometric and fluorescence polarisation studies'

D. Axelrod, (1979), Biophys. J., 26, 557-73;
'Carbocyanine dye orientation in red cell membrane studied by microscopic fluorescence polarisation'

A.J. Waring, P. Glatz and J.M. Vanderkooi, (1979), Biochim. Biophys. Acta, 557, 391-8;
'Subzero temperature study of the inner mitochondrial membrane and related phospholipid membrane systems with the fluorescent probe, trans-parinaric acid'

T. Ohyashiki, K. Chiba and T. Mohri, (1979), J. Biochem. (Tokyo), 86, 1479-85; Chem. Abstr., 92-2034;
'Terbium as a fluorescent probe for analysis of the nature of calcium ion-binding sites of rat intestinal mucosal membranes'

S. Shkenderov and K. Koburova, (1979), Izv. Durzh. Inst. Kontrol Lek. Sredstva, 12, 75-80; Chem. Abstr., 92-16754;

'Effect of some bee venom peptide components on the
fluorescent probe study of human serum albumin and human
erythrocyte membranes'

S. Wunderlich, F. Pliquett, V.A. Petrov, G.E. Dobrecov and
Y.A. Vladimirov, (1979), Wiss. Z. Karl-Marx-Univ., Leipzig,
Math. Naturwiss. Reihe, 28, 128-40; Chem. Abstr., 92-34945;
'Changes in phospholipid membrane induced by cholesterol'

O. Nagata, (1979), Kurume Igakkai Zasshi, 42, 579-602;
Chem. Abstr., 92-36152;
'Formation of lipid peroxide and injury of oxidative
phosphorylation in mitichondria by ultraviolet exposure.
Mitochondrial surface potential by fluorescence probes'

B. Rosdy, T. Kremmer, L. Holczinger and K. Bartha, (1979),
Proc. Hung. Annu. Meet. Biochem., 19th., 73-6; Chem. Abstr.
92-37119;
'Spectrofluorimetric studies on the binding of membrane-active agents to liver cell plasma membranes and to serum lipoproteins. 1. Procedures applied'

J. Michejda, J. Adamski, M. Hejnowicz and L. Hryniewiecka,
(1979), Dev. Bioenerg. Biomembr., 3 (Funct. Mol. Aspects
Biomembr. Trans.), 331-4; Chem. Abstr., 92-37436;
'Changes in membrane potential in amoeba mitochondria
monitored by the fluorescent probe diS-C3-(5)'

M. Krieger, L.C. Smith, R.G.W. Anderson, J.L. Goldstein,
Y.J. Kao, H.J. Pownall, A.M. Gotto and M.S. Brown, (1979),
J. Supramol. Struct., 10, 467-78;
'Reconstituted low density lipoprotein : a vehicle for the
delivery of hydrophobic fluorescent probes to cells'

C. Helene, J.J. Toulme and T.L. Doan, (1979), Nucleic Acids
Res., 7, 1945-54; Chem. Abstr., 92-71461;
'A spectroscopic probe for stacking interactions between
nucleic acid bases and tryptophan residues of proteins'

J.L. Browning and D.L. Nelson, (1979), J. Membr. Biol., 49,
75-103;
'Fluorescent probes for asymmetric lipid bilayers : synthesis
and properties in phosphatidyl choline liposomes and erythrocyte membranes'

A.B. Rawitch and R-Y. Hwan, (1979), Biochem. Biophys. Res.
Commun., 91, 1383-9;

K. Kano, A. Romero, B. Djermouni, H.J. Ache and J.H. Fendler, (1979), J. Amer. Chem. Soc., 101, 4030-7; 'Characterisation of surfactant vesicles as membrane mimetic viscosity. 2. Temperature-dependent changes of the turbidity, viscosity, fluorescence polarisation of 2-methylanthracene, and positron annihiliation in sonicated dioctadecyldimethyl-ammonium chloride'

J.M. Vanderkooi, (1979), Alcohol Clin. Exp. Res., 3, 60-3; Chem. Abstr., 91-103463; 'Effects of ethanol on membranes : a fluorescent probe study'

S.R. Johns, R.I. Willing, K.R. Thulborn and W.H. Sawyer, (1979), Chem. Phys. Lipids, 24, 11-16; 'Carbon-13 studies on fluorescent probes : carbon-13 chemical shifts and longitudinal relaxation times of N-hydroxy fatty (N = 2,6,9,12) acids and N-(9-anthroyloxy) stearic (N = 6, 12) acids'

J.C. Daniels, (1979), Lab. Res. Methods Biol. Med., 3, 199-209; Chem. Abstr., 91-86581; 'Comparison of fluid phase fluorescence and nephelometric analyses of serum proteins'

W. Schmidt, (1979), J. Membr. Biol., 47, 1-25; 'On the environment and the rotational motion of amphiphilic favins in artificial membrane vesicles as studied by fluorescence'

S. Takeuchi and A. Maeda, (1979), Biochim. Biophys. Acta, 563, 365-74; 'Fluorescein mercuric acetate as a probe of the dynamic structure of double-helical DNA'

K. Hildenbrand and C. Nicolau, (1979), Biochim. Biophys. Acta, 553, 365-77; 'Nanosecond fluorescence anisotropy decays of 1,6-diphenyl-1,3,5-hexatriene in membranes'

Y.G. Chu and C.R. Cantor, (1979), Nucleic Acids Res., 6, 2355-61; Chem. Abstr., 91-51425; 'Segmental flexibility in E.Coli ribosomal protein S1 as studied by fluorescence polarisation'

M.P. Heyn, (1979), FEBS Letters, 108, 359-64; 'Determination of lipid order parameters and rotational

correlation times from fluorescence depolarisation
experiments'

K. Kano and J.H. Fendler, (1979), Chem. Phys. Lipids,
23, 189-200;
'Dynamic fluorescence investigations of the effect of osmotic
shocks on the microenvironments of charged and uncharged
dipalmitoyl-D,L-α-phosphatidylcholine liposomes'

P.B. Fisher, M. Flamm, D. Schachter and I.B. Weinstein,
(1979), Biochem. Biophys. Res. Comm., 86, 1063-8;
'Tumor promoters induce membrane changes detected by
fluorescence polarisation'

J.K. Baird, E.T. Arakawa, D.W. Noid and H.R. Petty, (1979),
J. Chem. Phys., 71, 5081-9;
'Phase fluorometry as a probe of diffusion-controlled
molecular encounters in dense fluids'

S. Yonei, T. Todo and M. Kato, (1979), Radiat. Res., 80,
484-93;
'Evidence for a change in the fluidity of erythrocyte
membranes following x-irradiation by measurement of pyrene
excimer fluorescence'

K.R. Thulborn, L.M. Tilley, W.H. Sawyer and F.E. Treloar,
(1979), Biochim. Biophys. Acta, 558, 166-78;
'The use of N-(9-anthroyloxy) fatty acids to determine
fluidity and polarity gradients in phospholipid bilayers'

D.E. Koppel, (1979), Biophys. J., 28, 281-91;
'Fluorescence redistribution after photobleaching. A new
multipoint analysis of membrane translational dynamics'

F. Schneeweiss, D.V. Naquira, K. Rosenbeck and A.S. Schneider,
(1978 pub. 1979), Catecholamines : Basic Clin. Front., Proc.
Int. Catecholamine Symp., 4th., 1, 352-4; Chem. Abstr.
92-776;
'Fluorescent probe studies of isolated chromaffin cell
membranes'

D. Blakeslee, (1979), J. Natl. Cancer Inst., 63, 325-9;
'Mitogen-induced changes in the fluorescence polarisation
of fluorescein in normal human lymphocytes : a membrane
event ?'

B.G. Barisas and M.D. Leuther, (1979), Biophys. Chem.,
10, 221-9;
'Fluorescence photobleaching recovery measurement of protein
absolute diffusion constants'

A.J.W.G. Visser and A. Van Hoek, (1979), J. Biochem. Biophys.
Methods, 1, 195-208;
'The measurement of subnanosecond fluorescence decay of
flavins using time-correlated photon counting and a mode-
locked argon ion laser'

L.A. Sklar, G.P. Miljanich, S.L. Bursten and E.A. Dratz,
(1979), J. Biol. Chem., 254, 9583-91;
'Thermal lateral phase separations in bovine retinal rod
outer segment membranes and phospholipids as evidenced by
parinaric acid fluorescence polarisation and energy transfer'

D. Heldenberg, B. Werbin, L. Inbar, I. Tamir and M. Inbar,
(1979), Clin. Chem. Acta, 95, 493-500;
'Dynamics and composition of serum lipids in hyperlipo-
proteinemias'

R. Grienert, H. Staerk, A. Stier and A. Weller, (1979),
J. Biochem. Biophys. Methods, 1, 77-83;
'E-type delayed fluorescence depolarisation, a technique
to probe rotational motion in the microsecond range'

F. Hare, J. Amiell and C. Lussan, (1979), Biochim. Biophys.
Acta, 555, 388-408;
'Is an average viscosity tenable in lipid bilayers and
membranes ? A comparison of semi-empirical equivalent
viscosities given by unbound probes : a nitroxide and a
fluorophor'

P.I. Lelkes, A. Kapitkovsky, H. Eibl and I.R. Miller, (1979),
FEBS Letters, 103, 181-5;
'Headgroup-dependent modulation of phase transitions in
dipalmitoyl lecithin analogs. A fluorescence depolarisation
study'

C.S. Pike, J.A. Berry and J.K. Raison, (1979), Low Temp.
Stress Crop. Plants : Role Membr., (Proc. Int. Seminar),
305-18; Chem. Abstr., 93-41875;
'Fluorescence polarisation studies of membrane phospholipid
phase separations in warm and cool climate plants'

D. Georgescauld, J.P. Desmazes and H. Duclohier, (1979), Mol. Cell. Biochem., 27, 147-53; 'Temperature dependence of the fluorescence of pyrene-labelled crab nerve membrane'

M.J. Sinosich and T. Chard, (1979), Ann. Clin. Biochem., 16, 334-5; 'Fluoroimmunoassay of α-fetoprotein (AFP) in amniotic fluid'

I.H. Munro, I. Pecht and L. Stryer, (1979), Daresbury Lab. (Rep.), DL/SCI/R 1979, DL/SCI/R13, Appl. Synchrotron Radiat. Study Large Mol. Chem. Biol. Interest, 109-110; Chem. Abstr., 92-159190; 'Subnanosecond motions of tryptophan residues in proteins'

L. Letellier and E. Shechter, (1979), Eur. J. Biochem., 102, 441-7; 'Cyanine dye as monitor of membrane potentials in E. Coli cells and membrane vesicles'

J. Teissie, (1979), Chem. Phys. Lipids, 25, 357-68; 'A fluorescence study with polarised incident light of the compression of phospholipid bilayers spread at the air/water interface : orientation processes in the glycerol region'

J.C. Dederen, L. Coosemans, F.C. De Schryver and A. Van Dormael, (1979), Photochem. Photobiol., 30, 443-7; 'Complex solvent dependence of pyrenealdehyde fluorescence as a micellar polarity probe'

B. Mely-Goubert, F. Calvo and C. Rosenfield, (1979), Biomed. Express, 31, 155-6; 'Study of platelet membrane proteins through fluorescence polarisation of diphenylhexatriene'

F. Jaehnig, (1979), Proc. Natl. Acad. Sci. U.S.A., 76, 6361-5, 'Structural order of lipids and proteins in membranes: evaluation of fluorescence anisotropy data'

F. Jaehnig, (1979), J. Chem. Phys., 70, 3279-3290; 'Molecular theory of lipid membrane order'

J.R. Lakowicz, F.G. Prendergast and D. Hogen, (1979), Biochemistry, 18, 508-519; 'Differential polarised phase fluorometric investigations of diphenylhexatriene in lipid bilayers. Quantitation of hindered depolarising rotations'

'Anilinonaphthalenesulphonate as a probe for the native
structure of bovine alpha lactalbumin : absence of binding
to the native, monomeric protein'

L.M. Loew, S. Scully, L. Simpson and A.S. Waggoner, (1979),
Nature, 281, 497-9;
'Evidence for a charge-shift electrochromic mechanism in a
probe of membrane potential.

R.E. Hubbard and C.J. Garrett, (1979), Biochem. Soc. Trans.,
7, 993-5;
'Characterisation of the uptake of 1,6-diphenylhexa-1,3,5-
triene by adipocyte membranes'

G.E. Dobretsov, (1979), Itogi Nauki Tekh : Biofiz., 11, 101-
8; Chem. Abstr., 92-176616;
'Fluorescent probes : optical properties and membrane interactions'

A. Ceballos, P. Chaparro and J.R. Chantres Antoranz, (1979),
An. R. Acad. Farm., 45, 367-78; Chem. Abstr., 92-176784;
'Sodium salicylate as a fluorescent probe for measurements
of liposome permeability'

E. Carbone, Detect. Meas. Free Ca^{2+} cells, (1979), 355-71;
Elsevier, Amsterdam; Chem. Abstr., 92-177882;
'Aequorin and fluorescent chelating probes to detect free
calcium influx and membrane-associated calcium in excitable
cells'

J.M. Gonzalez-Ros, P. Calvo-Fernandez, V. Sator and
M. Martinez-Carrion, (1979), J. Supramol. Struct., 11, 327-
38;
'Pyrenesulphonyl azide as a fluorescent label for the study
of protein-lipid boundaries of acetylcholine receptors in
membranes'

M. Ehrenberg, R. Rigler and W. Wintermeyer, (1979),
Biochemistry, 18, 4588-95;
'On the structure and conformational dynamics of tRNA (Phe
yeast) in solution'

A. Gafni, (1980), Biochemistry, 19, 237-44;
'Acid-base equilibriums of the oxidised β-nicotinamide
adenine dinucleotide-pyruvate adduct in the ground and
electronically excited states. A proton transfer probe for
proteins'

L.A. Sklar, M.C. Doody, A.M. Gotto and H.J. Pownall, (1980), Biochemistry, 19, 1294-301; 'Serum lipoprotein structure : resonance energy transfer localisation of fluorescent lipid probes'

B. Mely-Goubert and M.H. Freedman, (1980), Cancer Biochem. Biophys., 4, 167-71; 'Membrane associated proteins and malignancy : an experimental hypothesis'

D.E. Wolf, M. Edidin and P.R. Dragston, (1980), Proc. Natl. Acad. Sci. U.S.A., 77, 2043-5; 'Effect of bleaching light on measurements of lateral diffusion in cell membranes by the fluorescent photobleaching recovery method'

A.M. Grigorova, N. Cittanova and G. Weber, (1980), Biochem. Biophys. Res. Comm., 94, 413-18; 'Existance of multiple sites for anilinonaphthalene sulphonate (ANS) in an α-fetoprotein fraction. Demonstration by fluorescence polarisation'

A.J.C. Fulford and W.E. Peel, (1980), Biochim. Biophys. Acta, 598, 237-46; 'Lateral pressures in biomembranes estimated from the dynamics of fluorescent probes'

P. Ghosh-Dastidar, D. Giblin, B. Yaghmai, A. Das, H.K. Das, L.J. Parkhurst and N.K. Gupta, (1980), J. Biol. Chem., 255, 3826-9; 'Protein synthesis in rabbit reticulocytes. 27. A study of the mechanism of interaction of fluorescently-labelled Co-eIF-2A with eIF-2 using fluorescence polarisation'

P. Baulding, P.A. Light and A.W. Preece, (1980), Protides Biol. Fluids, 27th, 435-8; Chem. Abstr., 93-5716; 'A study of lectin binding to mamalian lymphoid cells using fluorochromasia and fluorescence polarisation techniques'

J.A. Reidler, (1980), Diss. Abstr. Int. B, 40, 4094; 'Lateral diffusion and aggregation of fluorescent lipids and viral glycoproteins in cell plasma membranes'

G.D. Reinhart and H.A. Lardy, (1980), Biochemistry, 19, 1484-90; 'Rat liver phosphofructokinase : use of fluorescence polarisation to study aggregation at low protein concentration'

T.J. Rink, C. Motecucco, T.R. Hesketh and R.Y. Tsien, (1980),
Biochim. Biophys. Acta, 595, 15-30;
'Lymphocyte membrane potential assessed with fluorescent
probes'

J. Sunamoto, H. Kondo, T. Nomura and H. Okamato, (1980),
J. Amer. Chem. Soc., 102, 1146-52;
'Liposomal membranes. 2. Synthesis of a novel pyrene-labelled
lecithin and structural studies on liposomal bilayers'

R.B. Moreland and M.E. Dockter, (1980), Anal. Biochem., 103,
26-32;
'Preparation and characterisation of 3-azido-2,7-naphthalene
disulponate : a photolabile fluorescent precursor useful as
a hydrophilic surface probe'

G. Abraham and P.S. Low, (1980), Biochim. Biophys. Acta,
597, 285-91;
'Covalent labelling of specific membrane carbohydrate
residues with fluorescent probes'

E. Lavie and M. Sonenberg, (1980), FEBS Letters, 111, 281-4;
'Spectroscopic evidence for interactions of merocyanine 540
with valinomycin in the presence of potassium'

J.D. Johnson, M.R. Taskinen, N. Matsuoka and R.L. Jackson,
(1980), J. Biol. Chem., 255, 3461-5;
'Dansyl phosphatidylethanolamine-labelled very low density
lipoproteins. A fluorescent probe for monitoring lipolysis'

J.R. Lakowicz and D. Hogen, (1980), Chem. Phys. Lipids,
26, 1-40;
'Chlorinated hydrocarbon-cell membrane interactions studied
by the fluorescence quenching of carbazole-labelled phospholipids : probe synthesis and characterisation of the
quenching methodology'

E.P. Kharchenko, K.I. Shestak and A.A. Tkachenko, (1980),
Byull. Eksp. Biol. Med., 89, 307-9; Chem. Abstr., 93-40920;
'Probing the structure of bacterial deoxyribonucleoproteins
with exogenous and endogenous nucleases'

E.S. Tecoma and D. Wu, (1980), J. Bacteriol., 142, 931-8;
'Membrane deenergisation by colicin K affects fluorescence
of exogenously added by not biosynthetically esterified
parinaric acid probes in E. Coli'

D. Johnston and G. Melnykovych, (1980), Biochim. Biophys. Acta, 596, 320-4;
'Effects of dexamethasone on the fluorescence polarisation of diphenylhexatriene in hela cells'

J.R. Lakowicz, (1980), J. Biochem. Biophys. Methods, 2, 91-119;
'Fluorescence spectroscopic investigations of the dynamic properties of proteins, membranes and nucleic acids'

H.J. Galla and J. Luisetti, (1980), Biochim. Biophys. Acta, 596, 108-117;
Lateral and transversal diffusion and phase transitions in erythrocyte membranes. An excimer fluorescence study'

T.A. Brasitus, A.R. Tall and D. Schachter, (1980), Biochemistry, 19, 1256-61;
'Thermotropic transitions in rat intestinal plasma membranes studied by differential scanning calorimetry and fluorescence polarisation'

R.A. Laing, J. Fischbarg and B. Chance, (1980), Invest. Ophthalmol. Visual Sci., 19, 96-102;
'Noninvasive measurements of pyridine nucleotide fluorescence from the cornea'

J.R. Lakowicz and J.R. Knutson, (1980), Biochemistry, 19, 905-11;
'Hindered depolarising rotations of perylene in lipid bilayers. Detection of lifetime-resolved fluorescence anisotropy measurements'

J.R. Lakowicz and H. Cherek, (1980), J. Biol. Chem., 255, 831-4;
'Dipolar relaxation in proteins on the nanosecond time scale observed by wavelength-resolved phase fluorometry of tryptophan fluorescence'

J.W. Pettegrew, J.S. Nichols and R.M. Stewart, (1980), Ann. Neurol., 8, 381-6;
'Membrane studies in Huntingdon's disease : steady-state fluorescence studies of intact erythrocytes'

J.R. Lakowicz and G. Weber, (1980), Biophys. J., 32, 591-601;
'Nanosecond segmental mobilities of tryptophan residues in proteins observed by lifetime-resolved fluorescence anisotropies'

D. Scherman and J.P. Henry, (1980), Biochim. Biophys. Acta,
599, 150-66;
'Oxonol-U as a probe of chromaffin granule membrane potentials'

P. Lianos and G. Cremel, (1980), Photochem. Photobiol.,
31, 429-34;
'Environmental effects on the electronic spectral properties
of 1-pyrenecarboxaldehyde and their application in probing
biological structures'

C.S. Owen, (1980), J. Membr. Biol., 54, 13-20;
'A membrane-bound fluorescent probe to detect phospholipid
vesicle cell fusion'

R. Kraayenhof, (1980), Methods. Enzymol., 69 (Photosynth.
Nitrogen Fixation, pt. C), 510-20;
'Analysis of membrane architecture: fluorimetric approach'

D. Georgescauld, J.P. Desmasez, R. Lapouyade, A. Babeau,
H. Richard and M. Winnik, (1980), Photochem. Photobiol.,
31, 539-46;
'Intramolecular excimer fluorescence : a new probe of phase
transitions in synthetic phospolipid membranes'

A.S. Verkman, (1980), Biochim. Biophys. Acta, 599, 370-9;
'The quenching of an intramembrane fluorescent probe. A
method to study the binding and permeation of phloretin
through bilayers'

A. Kato and S. Nakai, (1980), Biochim. Biophys. Acta,
624, 13-20;
'Hydrophobicity determined by a fluorescence probe method
and its correlation with surface properties of proteins'

G.E. Dobretsov, A. Deev, A. Kosikov and Y.U. Vladimirov,
(198), Biofizika, 25, 763-4; Chem. Abstr. 93-163149;
'Fluorescence of the calcium ion-sensitive probe chlorotetracycline in membrane systems. II. Causes of calcium
ion dependent changes of fluorescence'

A.I. Deev, G.E. Dobretsov, A.I. Kosikov and Y.A. Vladimirov,
(1980), Biofizika, 25, 763; Chem. Abstr., 93-181151;
'Fluorescence in membrane systems with chlortetracycline
as a calcium ion sensitive probe. 1. Interaction with
phospholipid membranes'

B. Mely-Goubert and M.H. Freedman, (1980), Biochim. Biophys. Acta, 601, 315-27;
'Lipid fluidity and membrane protein monitoring using 1,6-diphenyl-1,3,5-hexatriene'

R.L. Hoover, D.K. Bhalla, S. Yanovich, M. Inbar and M.J. Karnovsky, (1980), J. Cell. Physiol., 103, 399-406;
'Effects of linoleic acid on capping, lectin mediated mitogenesis, surface antigen expression, and fluorescence polarisation in lymphocytes and BHK cells'

K.K. Onuki, T.Y. Kazue and M. Sukigara, (1980), Bull. Chem. Soc. Japan, 53, 1914-17;
'Effect of the phase transition in liposomes on the fluorescence amphiphilic cyanine dyes'

R. Fato, P. Ragaini, G. Lenaz and E. Bertoli, (1980), Boll. Soc. Ital. Biol. Sper., 56, 991-5; Chem. Abstr., 93-127600;
'Effect of some lipophilic substances on fluorescence polarisation of perylene in lipid vesicles and mitochondrial membranes'

K.A. Zachariasse, W. Kuenle and A. Weller, (1980), Chem. Phys. Lett., 73, 6-11;
'Intramolecular excimer fluorescence as a probe of fluidity changes and phase transitions in phosphatidylcholine bilayers'

G. Curatola, G. Lenaz, L. Mazzanti, G. Grilli and M. Familiari, (1980), Boll. Soc. Ital. Biol. Sper., 56, 527-32; Chem. Abstr., 93-125313;
'Effect of anesthetics on membrane fluidity tested by two different techniques : EPR spin labels and polarisation of perylene fluorescence'

G. Lipari and A. Szabo, (1980), Biophys. J., 30, 489-506;
'Effect of vibrational motion on fluorescence depolarisation and NMR relaxation in macromolecules and membranes'

T.L. Bushueva, E.P. Busel and E.A. Burshtein, (1980), Arch. Biochem. Biophys., 204, 161-6;
'Some regularities of dynamic accessibility of buried fluorescent residues to external quenchers in proteins'

P. Graceffa and S.S. Lehrer, (1980), J. Biol. Chem., 255, 296-300;

'The excimer fluorescence of pyrene-labelled tropomyosin.
A probe of conformational dynamics'

F.G. Herring, I. Tatischeff and G. Weeks, (1980), Biochim.
Biophys. Acta, 602, 1-9;
'The fluidity of plasma membranes of dictyostelium
discoideum. The effects of polyunsaturated fatty acid
incorporation assessed by fluorescence depolarisation and
electron paramagnetic resonance'

S.M. Sorcher, J.C. Bartholomew and M.P. Klein, (1980),
Biochim. Biophys. Acta, 610, 28-46;
'The use of fluorescence correlation spectroscopy to probe
chromatin in the cell nucleus'

H. Tanabe, K. Kurihara and Y. Kobatake, (1980), Biochemistry,
19, 5339-44;
'Changes in membrane potential and membrane fluidity in
Tetrahymena pyroformis in association with chemoreception
of hydrophobic stimuli : fluorescence studies'

B.N. Korvatovskii, G.P. Kukarskikh, V.B. Tusov, V.Z.
Pashchenko and L.B. Rubin, (1980), Dokl. Acad. Nauk SSSR,
253, 1251-5 (Biophys); Chem. Abstr., 93-235361;
'Picosecond fluorometry of pigment-protein complexes
enriched by photosystem I reactive centres'

R.J. Cherry, E.A. Nigg and G.S. Beddard, (1980), Proc.
Natl. Acad. Sci. U.S.A., 77, 5899-903;
'Oligosaccharide motion in erythrocyte membranes investigated by picosecond fluorescence polarisation and microsecond dichroism of an optical probe'

M. Almgren, (1980), J. Amer. Chem. Soc., 102, 7882-7;
'Migration and partitioning of pyrene and perylene between
lipid vesicles in aqueous solution studied with a
fluorescence-stopped flow technique'

R.D. Klausner and D.E. Wolf, (1980), Biochemistry, 19,
6199-203;
'Selectivity of fluorescent lipid analogues for lipid
domains'

Z. Derzko and K. Jacobson, (1980), Biochemistry, 19, 6050-7;
'Comparative lateral diffusion of fluorescent lipid
analogues in phospholipid multibilayers'

G.E. Dobretsov, U.Z. Lankin, T.A. Borshchevskaya, V.A. Petrov, V.N. Ivanov and N.V. Kotelevtseva, (1980), Byull. Eksp. Biol. Med., 90, 173-4; Chem. Abstr., 93-184763; 'Use of fluorescent probes for the detection of structural changes in liver endoplasmic reticulum membranes of rats fed an atherogenic diet'

L.A. Sklar and E.A. Dratz, (1980), FEBS Letters, 118, 308-10; 'Analysis of membrane bilayer asymmetry using parinaric acid fluorescent probes'

M.G.P. Vale and A.P. Carvalho, (1980), Biochim. Biophys. Acta, 601, 620-9; 'Interaction of chemical probes with sarcoplasmic reticulum membranes'

N. Gains, (1980), Eur. J. Biochem., 111, 199-202; 'The limitations of chlorotetracycline as a fluorescent probe of divalent cations associated with membranes'

T. Koop, (1980), Eesti NSU Tead. Akad. Toim., Biol., 29, 182-7; Chem. Abstr., 93-216017; 'Antioxidative properties of some fluorescent probes in lipid peroxidation in erythrocyte membranes'

S.M. Dunn, S.G. Blanchard and M.A. Raftery, (1980), Biochemistry, 19, 5645-52; 'Kinetics of carbamylcholine binding to membrane-bound acetylcholine receptor monitored by fluorescence changes of a covalently bound probe'

T. Koop, (1980), Radiobiologiya, 20, 648-53; Chem. Abstr., 94-1593; 'Use of fluorescent probes in the study of changes in membrane structures of cell nuclei during gamma-radiation-induced lipid peroxidation'

R.G. Ashcroft, K.R. Thulborn, J.R. Smith, H.G.L. Coster and W.H. Sawyer, (1980), Biochim. Biophys. Acta, 602, 299-308; 'Perturbations to lipid bilayers by spectroscopic probes as determined by dielectric measurements'

G.S. Jones, K. Van Dyke and V. Castranova, (1980), J. Cell. Physiol., 104, 425-31; 'Purification of human granulocytes by centrifugal elutriation and measurement of transmembrane potential'

D.D. Koblin, J. Yguerabide and H.H. Wang, (1980), Prog.
Anesthesiol., 2 (Mol. Mech. Anesth.), 439-46; Chem. Abstr.
94-10894;
'Interaction of local anesthetics with membrane proteins
and membrane lipids as studied by nanosecond and fluorescence spectroscopy'

J. Sunamoto, T. Nomura and H. Okamoto, (1980), Bull. Chem.
Soc. Japan, 53, 2768-72;
'Liposomal membranes. III. Permeation of pyrene-labelled
lecithin into matrix of liposomal bilayers'

K.Y. Law, (1980), Chem. Phys. Lett., 75, 545-9;
'Fluorescence probe for microenvironments : anomalous
viscosity dependence of the fluorescence quantum yield
of p-N,N-dialkylaminobenzylidenemalononitrile in 1-alkanols'

Z. Harel and C. Djerassi, (1980), Lipids, 15, 694-6;
'Dinosterol in model membranes : fluorescence polarisation
studies'

K. Fukuzawa, H. Chida and A. Suzuki, (1980), J. Nutr. Sci.
Vitaminol., 26, 427-34;
'Fluorescence depolarisation studies of phase transition
and fluidity in lecithin liposomes containing α-tocopherol'

F. Schroeder, (1980), Eur. J. Biochem., 112, 293-307,
'Fluorescent probes as monitors of surface membrane
fluidity gradients in murine fibroblasts'

M.A. O'Loughlin, D.W. Whillans and J.W. Hunt, (1980),
Radiat. Res., 84, 477-95;
'A fluorescent approach to testing the diffusion of oxygen
into mammalian cells'

J. Breton and N.E. Gaecintov, (1980), Biochim. Biophys.
Acta, 594, 1-32;
'Picosecond fluorescence kinetics and fast energy transfer
processes in photosynthetic membranes'

R. Greinert and A. Stier, (1980), Dev. Biochem., 13,
(Biochem. Biophys. Regul. Cytochrome P-450), 591-4;
Chem. Abstr., 94-78785;
'Rotational diffusion of cytochrome P-450 in a reconstit-
uted system measured by depolarisation of delayed
fluorescence'

J. Danner and H. Resnick, (198), Biochem. Pharmacol., 29, 2471-5;
'Use of the fluorescent probe, 1-anilino-8-naphthalene sulphonate, to monitor the interactions of chlorophenols with phospholipid membranes (liposomes)'

N. Kido, F. Tanaka, N. Kaneda and K. Yagi, (1980), Biochim. Biophys. Acta, 603, 255-65;
'Pulse fluorimetry of N-(1-pyrenesulphonyl) dipalmitoyl-L-α-phosphatidylethanolamine in concanavalin A-stimulated human lymphocytes'

R. Ford and J. Barber, (1980), Photobiochem. Photobiophys., 1, 263-70;
'The use of diphenylhexatriene to monitor the fluidity of the thylakoid membrane'

I.A. Bailey, C.J. Garratt. G.R. Penzer and D.S. Smith, (1980), FEBS Letters, 121, 246-8;
'The interaction of B29-fluoresceinthiocarbamyl-insulin with adipocyte membranes'

E.I. Dudich, I.V. Dudich and V.P. Timofeev, (1980), Mol. Immunol., 17, 1335-9;
'Fluorescence polarisation and spin-label study of human myeloma immunoglobulins A and M. Presence of segmental flexibility'

S. Kawato, S. Yoshida, Y. Orii, A. Ikegami and K. Kinosita, (1981), Biochim. Biophys. Acta, 634, 85-92;
'Nanosecond time-resolved fluorescence investigations of temperature-induced conformational changes in cytochrome oxidase in phosphatidylcholine vesicles and solubilised systems'

W. Herreman, P. Van Tornout, F.H. Van Cauwelaert and I. Hanssens, (1981), Biochim. Biophys. Acta, 640, 419-29;
'Interaction of α-lactalbumin with dimyristoyl phosphatidylcholine vesicles. II. A fluorescence polarisation study'

M. Deleers, J.M. Ruysschaert and W.J. Malaisse, (1981), Biochem. Biophys. Res. Comm., 98, 255-60;
'Glucose induces membrane changes detected by fluorescence polarisation in endocrine pancreatic cells'

L.M. Smith, H.M. McConnell, B.A. Smith and J.W. Parce,
(1981), Biophys. J., 33, 139-46;
'Pattern photobleaching of fluorescent lipid vesicles
using polarised laser light'

C-L. Wey, R.A. Cone and M.A. Edidin, (1981), Biophys. J.,
33, 225-32;
'Lateral diffusion of rhodopsin in photoreceptor cells
measured by fluorescence photobleaching and recovery'

E. Keh and B. Valeur, (1981), J. Colloid Interface. Sci.,
79, 465-78;
'Investigation of water-containing inverted micelles by
fluorescence polarisation. Determination of size and internal
fluidity'

V. Mikes amd J. Kovar, (1981), Biochim. Biophys. Acta,
640, 341-51;
'Interaction of liposomes with homologous series of
fluorescent berberine derivatives. New cationic probes for
measuring membrane potential'